A-Level Year 2

Biology

Exam Board: AQA

Everyone knows that life is complicated. As a Biology student, you'll know that life is *really* complicated... and you have to understand how it all works for the exams.

Never mind — at least this CGP book will make your revision a whole lot simpler. It has everything you'll need to master Year 2 of the AQA A-Level course, including crystal-clear study notes, realistic exam questions (with answers) and more!

There's also a free Online Edition you can read on your computer or tablet, so you don't even have to strain your myofibrils by carrying the book around.

How to access your free Online Edition

This book includes a free Online Edition to read on your PC, Mac or tablet. You'll just need to go to **cgpbooks.co.uk/extras** and enter this code:

0156 3155 7239 0437

By the way, this code only works for one person. If somebody else has used this book before you, they might have already claimed the Online Edition.

A-Level revision? It has to be CGP!

Published by CGP

From original material by Richard Parsons.

Editors:
Daniel Fielding, Andy Park, Rachael Rogers, Camilla Simson.

Contributors:
Sophie Anderson, Gloria Barnett, Jessica Egan, Mark Ellingham, Derek Harvey, Emily Lucas.

Cover image © duncan1890/iStockphoto.com

Data for graphs showing glucose concentration vs absorbance on pages 51 and 57 was obtained using a Mystrica colorimeter © Mystrica Ltd. www.mystrica.com

Data used to construct graph on page 81 showing the stock of spawning cod in the North Sea and the rate of mortality caused by fishing since 1960 from the International Council for the Exploration of the Sea.

Graph to show incidence of breast cancer defined by age and family history on page 88 reprinted from The Lancet, Volume 358, Issue No 9291. Familial Breast Cancer: collaborative reanalysis of individual data from 52 epidemiological studies including 58 209 women with breast cancer and 101 986 women without the disease, pgs 1389–1399 © Oct 2001 with permission from Elsevier.

Second graph on page 88 — Alcohol, tobacco & breast cancer — collaborative reanalysis of individual data from 53 epidemiological studies, including 58 515 women with breast cancer and 95 067 women without the disease. Reprinted by permission from Macmillan Publishers Ltd on behalf of Cancer Research UK: British Journal of Cancer © Nov 2002.

Data used to construct the table on the link between smoking and MDS on page 113 © 2013 Tong et al. A Meta-Analysis of the Relationship between Cigarette Smoking and Incidence of Myelodysplastic Syndromes. PLoS ONE 8(6): e67537. https://doi.org/10.1371/journal.pone.0067537. Reproduced under the terms of the Creative Commons Attribution License.

ISBN: 978 1 78908 025 4

With thanks to Philip Armstrong, Charlotte Burrows and Hayley Thompson for the proofreading.
With thanks to Jan Greenway for the copyright research.

Clipart from Corel®
Printed by Elanders Ltd, Newcastle upon Tyne.

Contents

Photosynthesis, Respiration and ATP

All organisms need energy for life processes (and you'll need some for revising), so it's pretty important stuff. Annoyingly, it's pretty complicated stuff too, but 'cos I'm feeling nice today we'll take it slowly, one bit at a time...

Biological Processes Need Energy

Plant and animal cells **need energy** for biological processes to occur:

- **Plants** need energy for things like **photosynthesis**, **active transport** (e.g. to take in minerals via their roots), **DNA replication**, **cell division** and **protein synthesis**.
- **Animals** need energy for things like **muscle contraction**, maintenance of **body temperature**, **active transport**, **DNA replication**, **cell division** and **protein synthesis**.

Without energy, these biological processes would stop and the plant or animal would die.

Photosynthesis Stores Energy in Glucose

1) **Photosynthesis** is the process where **energy** from **light** is used to **make glucose** from H_2O and CO_2 (the light energy is **converted** to **chemical energy** in the form of glucose).
2) Photosynthesis occurs in a **series** of **reactions**, but the overall equation is:

$$6CO_2 + 6H_2O + \text{Energy} \longrightarrow C_6H_{12}O_6 \text{ (glucose)} + 6O_2$$

3) So, energy is **stored** in the **glucose** until the plants **release** it by **respiration**.
4) Animals obtain glucose by **eating plants** (or **other animals**), then respire the glucose to release energy.

Cells Release Energy from Glucose by Respiration

1) **Plant** and **animal** cells **release energy** from **glucose** — this process is called **respiration**.
2) This energy is used to power all the **biological processes** in a cell.
3) There are two types of respiration:
   - **Aerobic respiration** — respiration **using oxygen**.
   - **Anaerobic respiration** — respiration **without oxygen**.
4) **Aerobic** respiration produces **carbon dioxide** and **water**, and releases **energy**. The overall equation is:

$$C_6H_{12}O_6 \text{ (glucose)} + 6O_2 \longrightarrow 6CO_2 + 6H_2O + \text{Energy}$$

5) **Anaerobic** respiration in **plants** and **yeast** produces **ethanol** and **carbon dioxide** and releases energy. In **humans**, anaerobic respiration produces **lactate** and releases energy.

ATP is the Immediate Source of Energy in a Cell

You should remember most of this stuff from **Topic 1**. Here's a quick recap:

1) A cell **can't** get its energy **directly** from glucose.
2) So, in respiration, the **energy released** from glucose is used to **make ATP** (adenosine triphosphate). ATP **carries energy** around the cell to where it's **needed**.

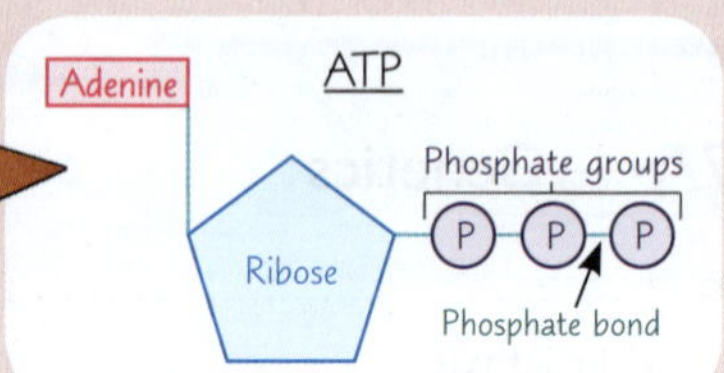

3) **ATP** is **synthesised** via a **condensation reaction** between **ADP** (adenosine diphosphate) and **inorganic phosphate** (P_i) using energy from an **energy-releasing** reaction, e.g. the **breakdown** of **glucose** in **respiration**. The energy is stored as **chemical energy** in the **phosphate bond**. The enzyme **ATP synthase** catalyses this reaction.
4) ATP **diffuses** to the part of the cell that **needs** energy.
5) Here, it's **hydrolysed** back into **ADP** and **inorganic phosphate** (P_i). Chemical **energy** is **released** from the phosphate bond and used by the cell. **ATP hydrolase** catalyses this reaction.
6) The ADP and inorganic phosphate are **recycled** and the process starts again.

Photosynthesis, Respiration and ATP

ATP has Specific Properties that Make it a Good Energy Source

1) ATP stores or releases only a **small**, **manageable amount** of energy at a time, so **no** energy is **wasted** as **heat**.
2) It's a **small**, **soluble** molecule so it can be **easily transported** around the cell.
3) It's **easily broken down**, so energy can be **easily released instantaneously**.
4) It can be **quickly re-made**.
5) It can make **other molecules** more **reactive** by **transferring** one of its **phosphate groups** to them (**phosphorylation**).
6) ATP **can't pass out** of the **cell**, so the cell **always** has an immediate supply of energy.

You Need to Know Some Basics Before You Start

There are some pretty confusing technical terms in this section that you need to get your head around:

- **Metabolic pathway** — a **series** of **small reactions** controlled by **enzymes**, e.g. **respiration** and **photosynthesis**.
- **Phosphorylation** — **adding phosphate** to a molecule, e.g. **ADP** is phosphorylated to **ATP** (see previous page).
- **Photophosphorylation** — **adding phosphate** to a molecule using **light**.
- **Photolysis** — the **splitting** (lysis) of a molecule using **light** (photo) energy.
- **Photoionisation** — when **light energy** '**excites**' **electrons** in an **atom** or **molecule**, giving them **more energy** and causing them to be **released**. The release of electrons causes the atom or molecule to become a **positively-charged ion**.
- **Hydrolysis** — the **splitting** (lysis) of a molecule using **water** (hydro).
- **Decarboxylation** — the **removal** of **carbon dioxide** from a molecule.
- **Dehydrogenation** — the **removal** of **hydrogen** from a molecule.
- **Redox reactions** — reactions that involve **oxidation** and **reduction**.

Remember redox reactions:

1) If something is **reduced** it has **gained electrons** (e^-), and may have **gained hydrogen** or lost oxygen.
2) If something is **oxidised** it has **lost electrons**, and may have **lost hydrogen** or gained oxygen.
3) Oxidation of one molecule **always** involves reduction of another molecule.

One way to remember electron and hydrogen movement is OILRIG. Oxidation Is Loss, Reduction Is Gain.

Photosynthesis and Respiration Involve Coenzymes

1) A **coenzyme** is a molecule that **aids** the **function** of an **enzyme**.
2) They work by **transferring** a **chemical group** from one molecule to another.
3) A coenzyme used in **photosynthesis** is **NADP**. NADP transfers **hydrogen** from one molecule to another — this means it can **reduce** (give hydrogen to) or **oxidise** (take hydrogen from) a molecule.
4) Examples of coenzymes used in **respiration** are: **NAD**, **coenzyme A** and **FAD**.
   - NAD and FAD transfer **hydrogen** from one molecule to another — this means they can **reduce** (give hydrogen to) or **oxidise** (take hydrogen from) a molecule.
   - **Coenzyme A** transfers **acetate** between molecules (see page 13).

When hydrogen is transferred between molecules, electrons are transferred too.

Warm-Up Questions

PRACTICE QUESTIONS

Q1 What is photoionisation?

Q2 Give the name of a coenzyme involved in photosynthesis.

Exam Question

Q1 ATP is the immediate source of energy inside a cell.
Describe how the synthesis and breakdown of ATP meets the energy needs of a cell. [6 marks]

Oh dear, I've used up all my energy on these two pages...

Well, I won't beat about the bush, this stuff is pretty tricky... nearly as hard as a cross between Hugh Jackman and concrete. With a little patience and perseverance (and plenty of [chocolate] [coffee] [marshmallows] — delete as you wish), you'll get there. Once you've got these pages straight in your head, the next ones will be easier to understand.

Photosynthesis

Right, pen at the ready. Check. Brain switched on. Check. Cuppa piping hot. Check. Sweets on standby. Check. Okay, I think you're all sorted to start photosynthesis. Finally, take a deep breath and here we go...

Photosynthesis Takes Place in the **Chloroplasts** of **Plant Cells**

1) **Chloroplasts** are **flattened organelles** surrounded by a **double membrane**. They are found in **plant cells**.
2) **Thylakoids** (fluid-filled sacs) are **stacked up** in the chloroplast into structures called **grana** (singular = **granum**). The grana are **linked** together by bits of thylakoid membrane called **lamellae** (singular = **lamella**).
3) Chloroplasts contain **photosynthetic pigments** (e.g. **chlorophyll a**, **chlorophyll b** and **carotene**). These are **coloured substances** that **absorb** the **light energy** needed for photosynthesis. The pigments are found in the **thylakoid membranes** — they're attached to **proteins**. The protein and pigment is called a **photosystem**.
4) There are **two** photosystems used by plants to capture light energy. **Photosystem I** (or PSI) absorbs light best at a wavelength of **700 nm** and **photosystem II** (PSII) absorbs light best at **680 nm**.
5) Contained within the inner membrane of the chloroplast and **surrounding** the thylakoids is a gel-like substance called the **stroma**. It contains **enzymes**, **sugars** and **organic acids**.
6) Carbohydrates produced by photosynthesis and not used straight away are stored as **starch grains** in the **stroma**.

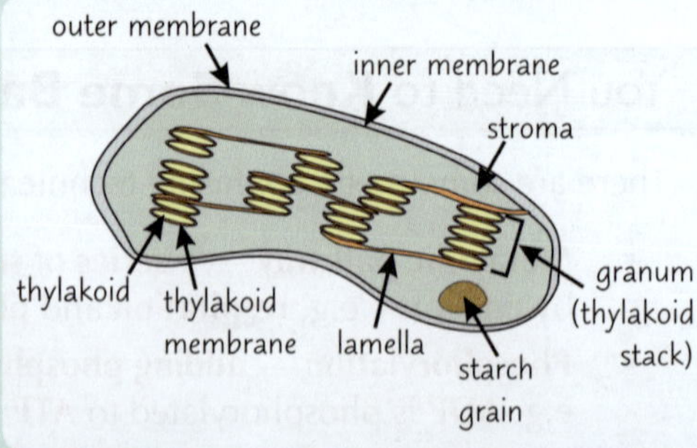

Photosynthesis can be Split into **Two Stages**

See p. 6 for loads more information on the Calvin cycle.

There are actually **two stages** that make up **photosynthesis**:

1 The Light-Dependent Reaction

1) As the name suggests, this reaction **needs light energy**.
2) It takes place in the **thylakoid membranes** of the chloroplasts.
3) Here, light energy is absorbed by **chlorophyll** (and other photosynthetic pigments) in the **photosystems**. The light energy **excites** the **electrons** in the **chlorophyll**, leading to their eventual **release** from the molecule. The **chlorophyll** has been **photoionised**.
4) Some of the **energy** from the **released electrons** is used to add a phosphate group to ADP to form **ATP**, and some is used to reduce NADP to form **reduced NADP**. ATP **transfers energy** and reduced NADP **transfers hydrogen** to the light-independent reaction.
5) During the process H_2O is **oxidised** to O_2.

2 The Light-Independent Reaction

1) This is also called the **Calvin cycle** and as the name suggests it **doesn't use light energy** directly. (But it does **rely** on the **products** of the light-dependent reaction.)
2) It takes place in the **stroma** of the chloroplast.
3) Here, the **ATP** and **reduced NADP** from the light-dependent reaction supply the **energy** and **hydrogen** to make **simple sugars** from CO_2.

This diagram shows how the two reactions link together in the chloroplast.

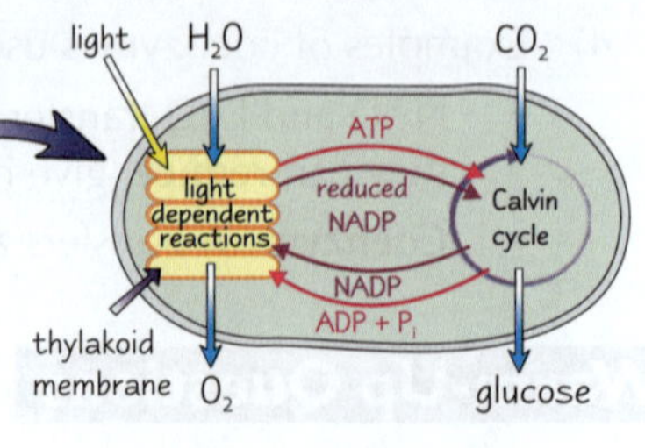

In the **Light-Dependent Reaction ATP** is Made by **Photophosphorylation**

In the light-dependent reaction, the **energy** resulting from the **photoionisation** of **chlorophyll** is used for three things:

1) Making **ATP** from **ADP** and **inorganic phosphate**. This reaction is called **photophosphorylation** (see p. 3).
2) Making **reduced NADP** from **NADP**.
3) Splitting **water** into **protons** (H^+ ions), **electrons** and **oxygen**. This is called **photolysis** (see p. 3).

The light-dependent reaction actually includes **two types** of **photophosphorylation** — **non-cyclic** and **cyclic**. Each of these processes has **different products** (see next page).

Photosynthesis

Non-cyclic Photophosphorylation Produces ATP, Reduced NADP and O_2

To understand the process you need to know that the photosystems (in the thylakoid membranes) are **linked** by **electron carriers**. Electron carriers are **proteins** that **transfer electrons**. The photosystems and electron carriers form an **electron transport chain** — a **chain** of **proteins** through which **excited electrons flow**. All the processes in the diagrams are happening together — I've just split them up to make it easier to understand.

1) Light energy excites electrons in chlorophyll

- **Light energy** is absorbed by **PSII**.
- The light energy **excites electrons** in **chlorophyll**.
- The electrons move to a **higher energy level** (i.e. they have more energy).
- These **high-energy electrons** are **released** from the **chlorophyll** and **move down** the **electron transport chain** to **PSI**.

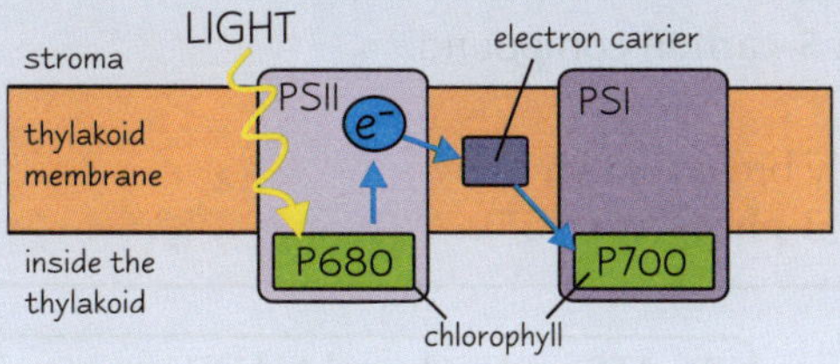

2) Photolysis of water produces protons (H^+ ions), electrons and O_2

- As the excited electrons **from chlorophyll leave PSII** to **move down** the electron transport chain, they must be **replaced**.
- **Light** energy splits **water** into **protons** (H^+ ions), **electrons** and **oxygen** — **photolysis**.
  (So the O_2 in photosynthesis comes from water and is made in the light-dependent reaction.)
- The reaction is: $H_2O \rightarrow 2H^+ + \frac{1}{2} O_2$

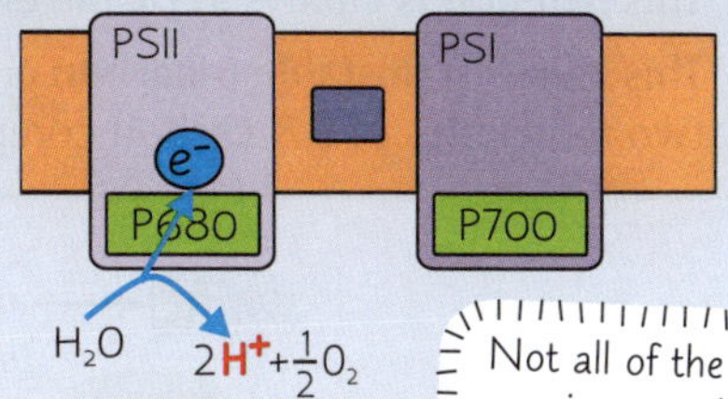

Not all of the electron carriers are shown in these diagrams.

3) Energy from the excited electrons makes ATP...

- The excited electrons **lose energy** as they **move down** the **electron transport chain**.
- This energy is used to **transport protons into** the **thylakoid**, so that the thylakoid has a **higher concentration** of protons than the stroma. This forms a **proton gradient** across the thylakoid membrane.
- Protons move **down** their concentration gradient, into the stroma, **via** the enzyme **ATP synthase**, which is **embedded** in the **thylakoid membrane**. The energy from this movement combines **ADP** and **inorganic phosphate** (P_i) to form **ATP**.

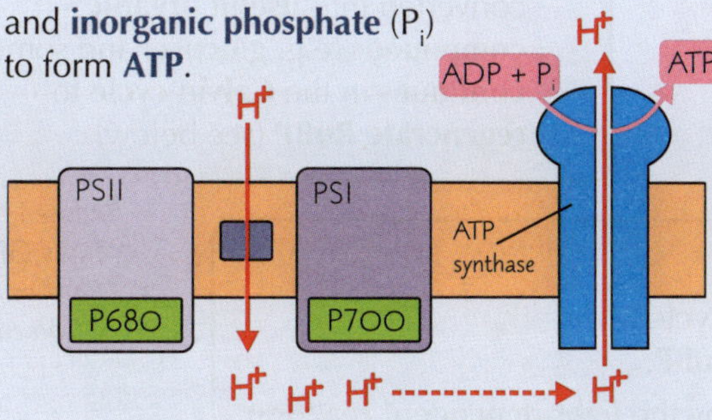

4) ...and generates reduced NADP.

- Light energy is **absorbed** by PSI, which excites the electrons again to an **even higher** energy level.
- Finally, the electrons are **transferred** to **NADP**, along with a **proton** (H^+ ion) from the **stroma**, to form **reduced NADP**.

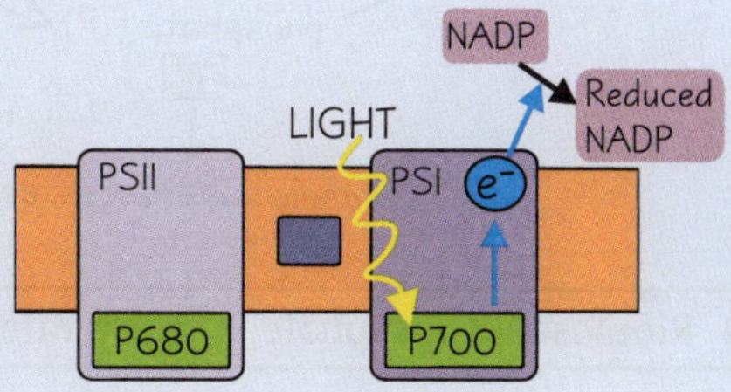

Remember a 'proton' is just another word for a hydrogen ion (H^+).

The process of **electrons** flowing down the **electron transport chain** and creating a **proton gradient** across the **membrane** to drive **ATP synthesis** is called **chemiosmosis**. It's described by the **chemiosmotic theory**.

Cyclic Photophosphorylation Only Produces ATP

Cyclic photophosphorylation **only uses PSI**. It's called 'cyclic' because the electrons from the chlorophyll molecule **aren't** passed onto NADP, but are **passed back** to PSI via electron carriers. This means the electrons are **recycled** and can repeatedly flow through PSI. This process doesn't produce any reduced NADP or O_2 — it **only produces** small amounts of **ATP**.

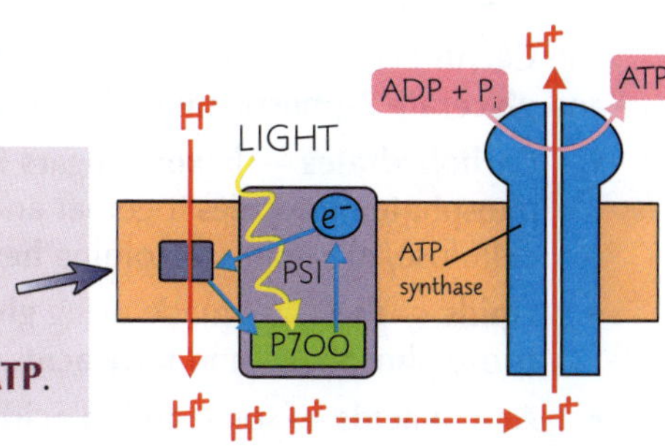

Photosynthesis

Don't worry, you're over the worst of photosynthesis now. Instead of electrons flying around, there's a nice cycle of reactions to learn. What more could you want from life? Money, fast cars and nice clothes have nothing on this...

The **Light-Independent** Reaction is also called the **Calvin Cycle**

1) The **Calvin cycle** takes place in the **stroma** of the chloroplasts.
2) It makes a molecule called **triose phosphate** from CO_2 and **ribulose bisphosphate** (a 5-carbon compound). Triose phosphate can be used to make **glucose** and other **useful organic substances** (see below).
3) There are a few steps in the cycle, and it needs **ATP** and **H^+ ions** to keep it going.
4) The reactions are linked in a **cycle**, which means the starting compound, **ribulose bisphosphate**, is **regenerated**.

The Calvin cycle is also known as carbon dioxide fixation because carbon from CO_2 is 'fixed' into an organic molecule.

Here's what happens at each stage in the cycle:

1) Carbon dioxide is combined with ribulose bisphosphate to form two molecules of glycerate 3-phosphate

- CO_2 enters the leaf through the **stomata** and diffuses into the **stroma** of the chloroplast.
- Here, it's combined with **ribulose bisphosphate (RuBP)**, a **5-carbon** compound. This reaction is catalysed by the enzyme **rubisco**.
- This gives an **unstable 6-carbon** compound, which quickly breaks down into **two** molecules of a **3-carbon** compound called **glycerate 3-phosphate (GP)**.

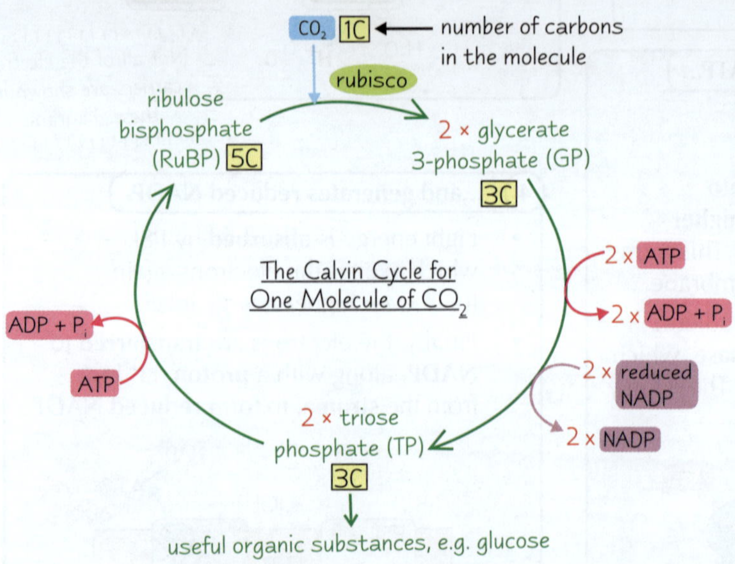

2) ATP and reduced NADP are required for the reduction of GP to triose phosphate

- The hydrolysis of **ATP** (from the **light-dependent** reaction) **provides energy** to turn the **3-carbon** compound, **GP**, into a **different** 3-carbon compound called **triose phosphate (TP)**.
- This reaction also requires **H^+ ions**, which come from **reduced NADP** (also from the **light-dependent reaction**). **Reduced NADP** is **recycled** to **NADP**.
- Some **triose phosphate** is then converted into **useful organic compounds** (e.g. **glucose**) and some **continues** in the **Calvin cycle** to **regenerate RuBP** (see below).

Reduced NADP reduces GP to TP — reduction reactions are explained on p. 3.

3) Ribulose bisphosphate is regenerated

- **Five** out of every **six** molecules of **TP** produced in the cycle aren't used to make hexose sugars, but to **regenerate RuBP**.
- Regenerating RuBP uses the **rest** of the **ATP** produced by the **light-dependent reaction**.

TP and **GP** are **Converted** into **Useful Organic Substances** like **Glucose**

The **Calvin cycle** is the starting point for making **all** the organic substances a plant needs. **Triose phosphate** (TP) and **glycerate 3-phosphate** (GP) molecules are used to make **carbohydrates**, **lipids** and **amino acids**:

- **Carbohydrates** — **hexose sugars** (e.g. glucose) are made by joining **two triose phosphate molecules** together and **larger** carbohydrates (e.g. sucrose, starch, cellulose) are made by joining **hexose sugars** together in **different ways**.
- **Lipids** — these are made using **glycerol**, which is synthesised from **triose phosphate**, and **fatty acids**, which are synthesised from **glycerate 3-phosphate**.
- **Amino acids** — some **amino acids** are made from **glycerate 3-phosphate**.

Hexose sugars are simple six carbon sugars.

Photosynthesis

The Calvin Cycle Needs to Turn Six Times to Make One Hexose Sugar

Here's the reason why:

1) **Three turns** of the cycle produces **six** molecules of **triose phosphate** (TP), because two molecules of TP are made for every one CO_2 molecule used.
2) **Five** out of **six** of these TP molecules are used to **regenerate ribulose bisphosphate** (RuBP).
3) This means that for **three turns** of the cycle only **one TP** is produced that's used to make a **hexose sugar**.
4) A hexose sugar has **six carbons** though, so **two TP** molecules are needed to form one hexose sugar.
5) This means the cycle must turn **six times** to produce **two molecules** of **TP** that can be used to make **one hexose sugar**.
6) Six turns of the cycle need **18 ATP** and **12 reduced NADP** from the light-dependent reaction.

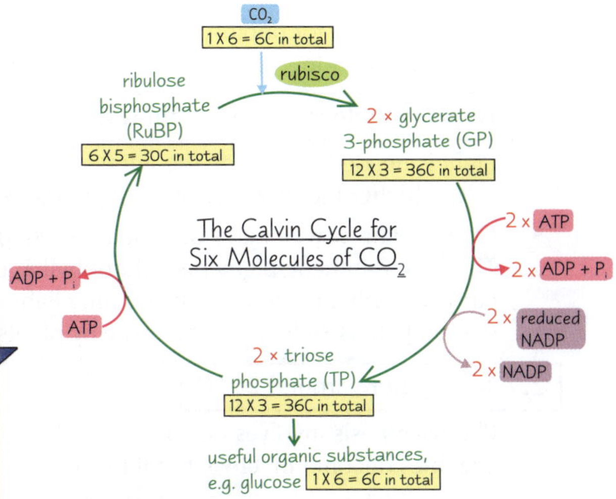

This might seem a bit inefficient, but it keeps the cycle going and makes sure there's always **enough RuBP** ready to combine with CO_2 taken in from the atmosphere.

Morag had to turn one million times to make a sock... two million for a scarf.

Warm-Up Questions

Q1 Name two photosynthetic pigments in the chloroplasts of plants.

Q2 At what wavelength does photosystem I absorb light best?

Q3 What three substances does non-cyclic photophosphorylation produce?

Q4 Which photosystem is involved in cyclic photophosphorylation?

Q5 Where in the chloroplasts does the light-independent reaction occur?

Q6 How many carbon atoms are there in a molecule of TP?

Q7 Name two organic substances made from triose phosphate.

Q8 How many CO_2 molecules need to enter the Calvin cycle to make one hexose sugar?

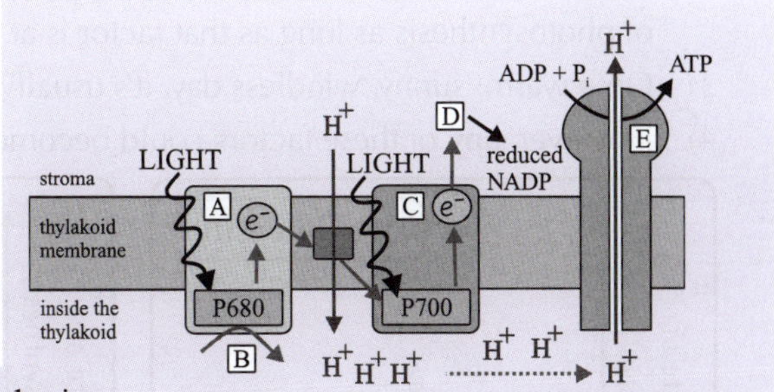

Exam Questions

Q1 The diagram above shows the light-dependent reaction of photosynthesis.

a) What does object A represent? [1 mark]

b) Describe process B and explain its purpose. [3 marks]

c) Explain how reactant D is made into reduced NADP. [2 marks]

Q2 Rubisco is an enzyme that catalyses the first reaction of the Calvin cycle. CA1P is an inhibitor of rubisco.

a) Describe how triose phosphate is produced in the Calvin cycle. [5 marks]

b) Briefly explain how ribulose bisphosphate (RuBP) is regenerated in the Calvin cycle. [2 marks]

c) Explain the effect that CA1P would have on glucose production. [3 marks]

Calvin cycles — bikes made by people who normally make pants...

Next thing we know there'll be people swanning about in their pants riding highly fashionable bikes. Sounds awful I know, but let's face it, anything would look better than cycling shorts. Anyway, it would be a good idea to go over these pages a couple of times — I promise you, there's still room left in your head for more information.

Limiting Factors in Photosynthesis

I'd love to tell you that you'd finished photosynthesis... but I'd be lying.

There are **Optimum Conditions** for **Photosynthesis**

The **ideal conditions** for photosynthesis vary from one plant species to another, but the conditions below would be ideal for **most** plant species in temperate climates like the UK.

1. High light intensity of a certain wavelength

- Light is needed to provide the **energy** for the **light-dependent reaction** — the **higher** the **intensity** of the light, the **more energy** it provides.
- Only certain **wavelengths** of light are used for photosynthesis. The photosynthetic pigments chlorophyll a, chlorophyll b and carotene only **absorb** the **red** and **blue** light in sunlight. (**Green** light is **reflected**, which is why plants look green.)

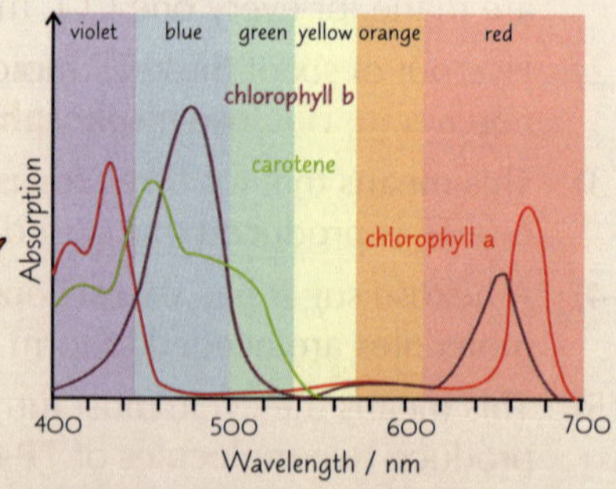

2. Temperature around 25 °C

- Photosynthesis involves **enzymes** (e.g. ATP synthase, rubisco). If the temperature falls **below 10 °C** the enzymes become **inactive**, but if the temperature is **more than 45 °C** they may start to **denature**.
- Also, at **high** temperatures **stomata close** to avoid losing too much water. This causes photosynthesis to slow down because **less CO_2** enters the leaf when the stomata are closed.

3. Carbon dioxide at 0.4%

- Carbon dioxide makes up **0.04%** of the gases in the atmosphere.
- Increasing this to **0.4%** gives a **higher rate** of photosynthesis, but any higher and the stomata start to **close**.

Plants also need a **constant supply** of **water** — **too little** and photosynthesis has to **stop** but **too much** and the soil becomes **waterlogged** (**reducing** the uptake of **minerals** such as **magnesium**, which is needed to make **chlorophyll a**).

Light, **Temperature** and **CO_2** can all **Limit Photosynthesis**

1) **All three** of these things need to be at the **right level** to allow a plant to photosynthesise as quickly as possible.
2) If any **one** of these factors is **too low** or **too high**, it will **limit photosynthesis** (slow it down). Even if the other two factors are at the perfect level, it won't make **any difference** to the speed of photosynthesis as long as that factor is at the wrong level.
3) On a warm, sunny, windless day, it's usually **CO_2** that's the limiting factor, and at night it's the **light intensity**.
4) However, **any** of these factors could become the limiting factor, depending on the **environmental conditions**.

Light intensity

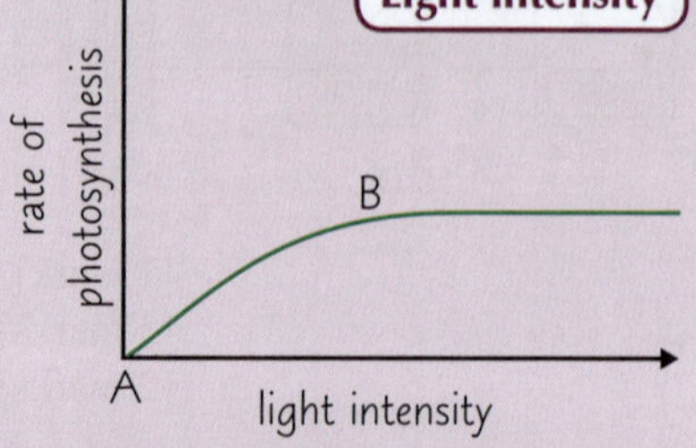

Between points A and B, the rate of photosynthesis is limited by the **light intensity**. So as the light intensity **increases**, so can the rate of photosynthesis. Point B is the **saturation point** — increasing light intensity after this point makes no difference, because **something else** has become the limiting factor. The graph now **levels off**.

Temperature

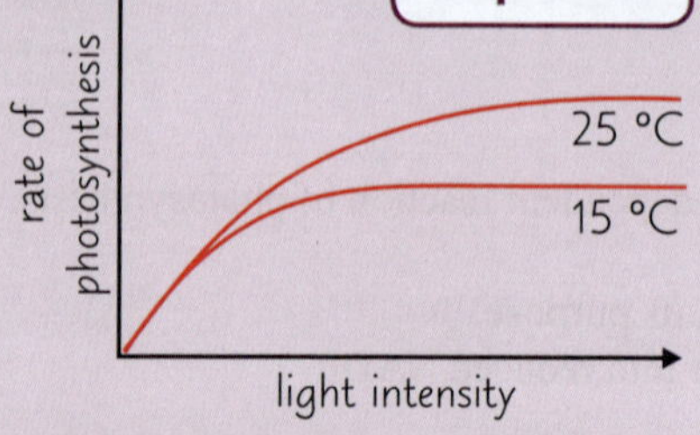

Both these graphs level off when **light intensity** is no longer the limiting factor. The graph at **25 °C** levels off at a **higher point** than the one at **15 °C**, showing that **temperature** must have been a limiting factor at **15 °C**.

The saturation point is where a factor is no longer limiting the reaction — something else has become the limiting factor.

CO_2 Concentration

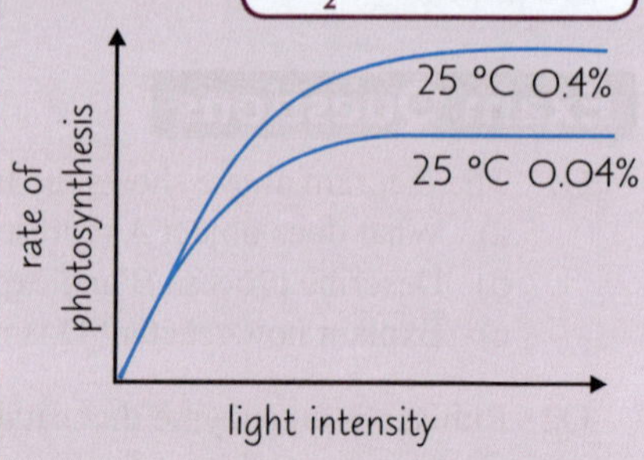

Again, both these graphs level off when **light intensity** is no longer the limiting factor. The graph at **0.4% CO_2** levels off at a **higher point** than the one at **0.04%**, so **CO_2 concentration** must have been a limiting factor at **0.04% CO_2**. The limiting factor here **isn't temperature** because it's the **same** for both graphs (25 °C).

Limiting Factors in Photosynthesis

Growers Use Information About Limiting Factors to Increase Plant Growth

Agricultural growers (e.g. farmers) know the **factors** that **limit photosynthesis** and therefore limit **plant growth**. This means they try to create an **environment** where plants get the **right amount** of everything that they need, which **increases growth** and so **increases yield**.

Growers create optimum conditions in **glasshouses** in the following ways:

Limiting Factor	Management in Glasshouse
Carbon dioxide concentration	CO_2 is added to the air, e.g. by burning a small amount of propane in a CO_2 generator.
Light	Light can get in through the glass. Lamps provide light at night-time.
Temperature	Glasshouses trap heat energy from sunlight, which warms the air. Heaters and cooling systems can also be used to keep a constant optimum temperature, and air circulation systems make sure the temperature is even throughout the glasshouse.

Similar techniques can also be used in **polytunnels** (tunnels made of polythene, under which plants can be grown).

You Need to be Able to Interpret Data on Limiting Factors

Here are some **examples** of the kind of **data** you might get in the exam:

The graph on the **right** shows the effect on plant growth of **adding carbon dioxide** to a greenhouse.

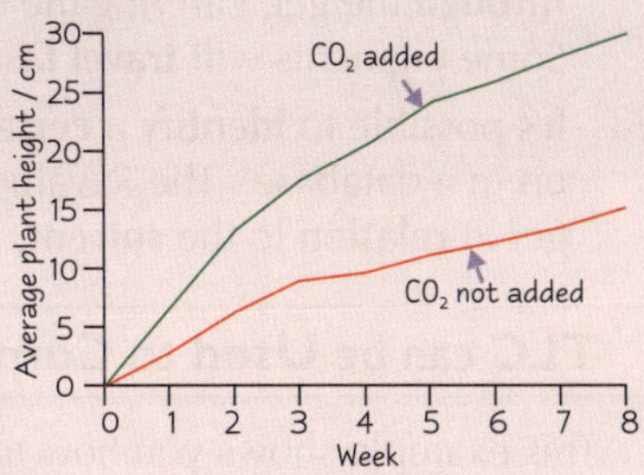

1) In the greenhouse **with added CO_2** plant **growth** was **faster** (the line is steeper) and on average the plants were **larger** after 8 weeks than they were in the control greenhouse (30 cm compared to only 15 cm in the greenhouse where no CO_2 was added).
2) This is because the plants use CO_2 to produce **glucose** by photosynthesis. The more CO_2 they have, the more glucose they can produce, meaning they can **respire more** and so have **more ATP** for **DNA replication**, **cell division** and **protein synthesis**, i.e. growth.

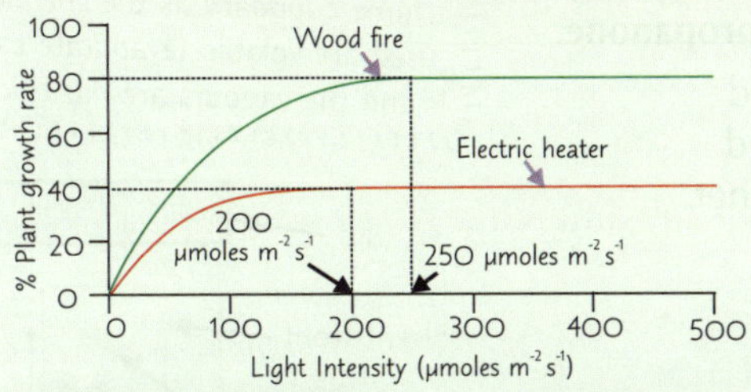

The graph on the **left** shows the effect of **light intensity** on plant growth, and the effect of two **different types** of **heater**.

1) At the start of the graph, the **greater** the **light intensity** the **greater** the **plant growth**.
2) At **200 μmoles m^{-2} s^{-1}** (micromoles per metre^2 per second) of light the **bottom** graph flattens out, showing that **CO_2 concentration** or **temperature** is **limiting growth** in these plants.
3) At **250 μmoles m^{-2} s^{-1}** of light the **top** graph flattens out. The difference between the two graphs could be because the **wood fire increases** the **temperature more** than the electric heater or because it's **increasing** the **concentration** of **CO_2** in the air (an electric heater **doesn't** release CO_2).

Warm-Up Questions

Q1 Name two factors that can limit plant growth.

PRACTICE QUESTIONS

Exam Question

Crop	Yield in glasshouse / kg	Yield grown outdoors / kg
Tomato	1000	200
Lettuce	750	230
Potato	850	680
Wheat	780	550

Q1 The table above shows the yields of various crops when they are grown in glasshouses and when grown outdoors.

a) Yields are usually higher overall in glasshouses.
Describe two ways in which conditions can be controlled in glasshouses to increase yields. [2 marks]

b) Glasshouses are not always financially viable for all crops.
Which crop above benefits the least from being grown in glasshouses? Explain your answer. [2 marks]

I'm a whizz at the factors that limit revision...

... watching Hollyoaks, making tea, watching EastEnders, walking the dog... not to mention staring into space (one of my favourites). Anyway, an interpreting data question could well come up in the exams — it could be any kind of data, but don't panic if it's not like the graphs above — as long as you understand limiting factors you'll be able to interpret it.

Photosynthesis Experiments

PRACTICAL SKILLS

Everyone loves a good experiment — especially when they involve bright colours. Here are two really colourful photosynthesis experiments for you to enjoy. Let's chop up some plants and marvel at the beauty of Biology...

You Can **Investigate** the **Pigments** in **Leaves** Using **Chromatography**

1) All plants contain several **different photosynthetic pigments** in their leaves. Each pigment absorbs a **different wavelength** of light, so having **more than one** type of pigment **increases** the **range** of **wavelengths** of light that a plant can **absorb**.
2) In addition to photosynthetic pigments, some plants also have **other pigments** in their leaves, which play other essential roles, e.g. protecting the leaves from excessive UV radiation. **Different species** of plants contain **different proportions** and **mixtures** of **pigments**.
3) You can use **thin layer chromatography** (**TLC**) to determine what **pigments** are present in the leaves of a plant. Like all chromatography, TLC involves:
   - A **mobile phase** — where molecules can move. In TLC, this is a **liquid solvent**.
   - A **stationary phase** — where molecules can't move. In TLC, this consists of a **solid** (e.g. glass) **plate** with a **thin layer of gel** (e.g. silica gel) on top.
4) A **sample** of pigments can be **extracted** from the plant and put on the TLC plate. When the plate is placed vertically in the **solvent**, the solvent moves upwards **through** the gel, carrying the dissolved pigments with it. Some pigments will **travel faster** or **further** through the gel than others, which **separates** them out.
5) It's possible to **identify** a **certain pigment** by calculating its **R_f value** and looking it up in a database. The R_f value is the **distance** a substance has moved through the gel in **relation** to the **solvent**. Each pigment has a specific R_f value.

TLC can be **Used** to **Compare** the **Pigments** in **Different Plants**

This example shows you how to use **TLC** to **compare** the **pigments** present in **shade-tolerant plants** and **shade-intolerant plants**. Make sure you're wearing a lab coat, eye protection and gloves before you start. Many of the chemicals involved are toxic and highly flammable.

1) **Grind** up several leaves from the shade-tolerant plant you're investigating with some **anhydrous sodium sulfate**, then add a few drops of **propanone**.
2) **Transfer** the **liquid** to a test tube, add some **petroleum ether** and gently shake the tube. **Two distinct layers** will form in the liquid — the **top layer** is the **pigments** mixed in with the petroleum ether.
3) Transfer some of the liquid from the **top layer** into a second test tube with some **anhydrous sodium sulfate**.
4) Draw a horizontal **pencil line** near the bottom of a **TLC plate**. Build up a single **concentrated spot** of the liquid from step 3 on the line by applying several drops and ensuring each one is **dry** before the next is added. This is the **point of origin**.
5) Once the point of origin is completely dry, put the plate into a small glass container with some prepared **solvent** (e.g. a mixture of **propanone**, **cyclohexane** and **petroleum ether**) — just enough so that the **point of origin** is a little bit **above** the solvent. Put a **lid** on the container and leave the plate to develop. As the solvent spreads up the plate, the different **pigments** move with it, but at **different rates** — so they **separate**.
6) When the solvent has **nearly** reached the top, take the plate out and **mark** the **solvent front** (the furthest point the solvent has reached) with a **pencil** and leave the plate to dry in a well-ventilated place.
7) There should be **several** new coloured spots on the chromatography plate between the **point of origin** and the **solvent front**. These are the separated **pigments**. You can calculate their **R_f values** and look them up in a database to **identify** what the pigments are.
8) **Repeat** the process for the **shade-intolerant** plant you're investigating and **compare** the **pigments** present in their leaves.

It's best to do steps 2 and 5 in a fume cupboard as the chemicals used are volatile (evaporate easily) and the vapours are hazardous.

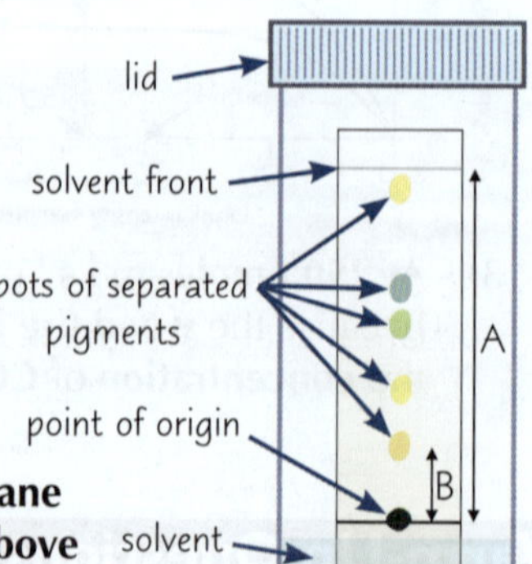

$$R_f \text{ value} = \frac{B}{A} = \frac{\text{distance travelled by spot}}{\text{distance travelled by solvent}}$$

You may find that the mixture of pigments in the leaves of the shade-tolerant plant is quite different compared to the shade-intolerant plant. One way that shade-tolerant plants can **adapt** to the **light conditions** in their environment is by possessing a **different proportion** of photosynthetic pigments, which allows the plant to make the best use of the light available to it. The mixture of non-photosynthetic pigments is also likely to be different. For example, the chloroplasts of shade-tolerant plants are adapted for photosynthesis in low light conditions, but really sensitive to higher levels of light. These plants sometimes produce **dark red** and **purple pigments** called **anthocyanins**, which are thought to **protect** their **chloroplasts** from **brief exposure** to **higher** light levels.

Photosynthesis Experiments

PRACTICAL SKILLS

You Can **Investigate** the **Activity** of **Dehydrogenase** in **Chloroplasts**

1) In **photosystem I**, during the **light-dependent** stage of photosynthesis, **NADP** acts as an **electron acceptor** and is **reduced** (see page 5). The reaction is **catalysed** by a **dehydrogenase enzyme**.
2) The **activity** of this **enzyme** can be investigated by adding a **redox indicator dye** to extracts of **chloroplasts**. Like **NADP**, the dye acts as an **electron acceptor** and gets **reduced** by the dehydrogenase in the chloroplasts. As the dye gets reduced, you'll see a **colour change**. For example, the dye **DCPIP** changes from **blue** to **colourless** when it gets reduced.
3) You can measure the rate of the dehydrogenase activity by measuring the **rate** at which DCPIP **loses** its **blue colour**. To do this, you need a **colorimeter**. A colorimeter measures how much light a solution **absorbs** when a light source is **shone** directly **through** it. A **coloured solution** absorbs **more light** than a colourless solution.

The experiment below shows you how to **investigate** the **effect** of **light intensity** on **dehydrogenase activity** in extracts of chloroplasts. It uses a **bench lamp** as a **light source** and involves placing tubes of chloroplast extract mixed with DCPIP at a range of **different distances** from the light source. **Light intensity** should **decrease** with **increasing distance** from the lamp. You'll need to **choose** the **distances** you're going to investigate (e.g. 15 cm, 30 cm and 45 cm) before you start.

You can use a similar method to investigate the effects of other factors on dehydrogenase activity in chloroplasts.

1) **Cut** a few **leaves** (spinach works well) into pieces. Remove any tough stalks.
2) Using a pestle and mortar, **grind** up the **leaf pieces** with some **chilled isolation solution** (a solution of **sucrose**, **potassium chloride** and **phosphate buffer** at **pH 7**). **Filter** the **liquid** you make into a **beaker** through a **funnel** lined with **muslin cloth**.
3) Transfer the liquid to **centrifuge tubes** and centrifuge them at **high speed** for **10 minutes**. This will make the **chloroplasts** gather at the **bottom** of each tube in a **'pellet'**.
4) **Get rid** of the **liquid** from the top of the tubes, **leaving** the **pellets** in the bottom.
5) **Re-suspend** the pellets in **fresh**, chilled **isolation solution**. This is your **chloroplast extract**. **Store** it on **ice** for the rest of the experiment.
6) Set up a **colorimeter** with a **red filter** and **zero** it using a cuvette (a cuboid-shaped vessel used in colorimeters) containing the **chloroplast extract** and **distilled water**.
7) Set up a **test tube rack** at a **set distance** from a **bench lamp**. Switch the lamp on.
8) Put a test tube in the rack, add a **set volume** of **chloroplast extract** to the tube and a **set volume** of **DCPIP**. **Mix** the contents of the tube together.
9) **Immediately** take a sample of the mixture from the tube and add it to a **clean cuvette**. Then place the cuvette in your colorimeter and **record** the **absorbance**. Do this every **2 minutes** for the next **ten minutes**.
10) **Repeat** steps **7** to **9** for **each distance** under investigation.

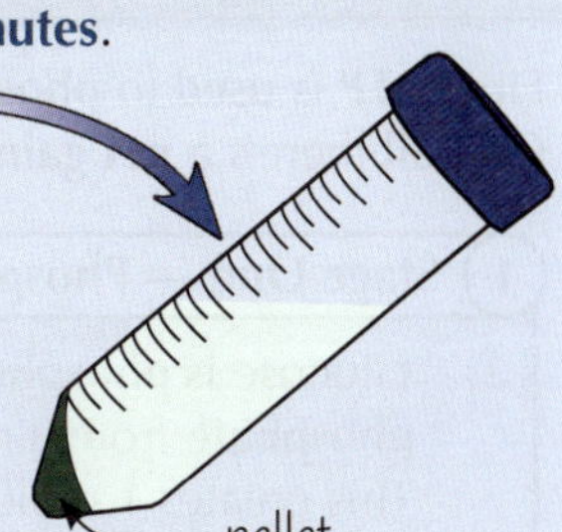

If dehydrogenase activity is taking place, the **absorbance** will **decrease** as the DCPIP gets **reduced** and **loses** its **blue colour**. The **faster** the absorbance decreases, the **faster** the **rate** of **dehydrogenase activity**. You can plot a **graph** of **absorbance against time** for each distance from the light source. Then **compare** your results to **determine** how light intensity affects the **rate** of the dehydrogenase enzyme.

You should also check whether the absorbance changes at each distance in two negative control tubes. The first should contain only DCPIP and chilled isolation solution (no chloroplast extract). The second should contain both DCPIP and chloroplast extract, but it should be wrapped in tin foil (so no light reaches the contents of the tube). No change in absorbance should be seen for these two controls.

Warm-Up Questions

PRACTICE QUESTIONS

Q1 What is a chromatography plate?

Exam Question

Q1 A group of scientists was interested in how light intensity can affect the activity of dehydrogenase enzymes in photosynthesis. They prepared a sample of isolated chloroplasts and added a redox indicator dye. They then used a colorimeter to measure the absorption of the solution at regular intervals when placed in different light intensities.

a) What role do dehydrogenase enzymes play in photosystem I of photosynthesis? [1 mark]

b) Explain how a redox indicator dye is able to indicate dehydrogenase activity in photosystem I. [3 marks]

No animals were harmed in the making of these experiments...

... but I did ruin my garden. Make sure that the plant's sacrifice wasn't in vain and learn how these experiments work. You might get a question in your exams that involves experiments pretty similar to these, so they're worth remembering.

Respiration

Roses are red, violets are blue, I love respiring and I bet you do too. Now you've enjoyed that poem, it's time to concentrate. I hope you like remembering reactions involved in respiration, because these pages have several.

There are Two Types of Respiration

1) The two types of respiration are **aerobic** (requires oxygen) and **anaerobic** (doesn't require oxygen).
2) Both **produce ATP** (see p. 2), although **anaerobic respiration** produces **less**.
3) Both **start** with the process of **glycolysis** (see below). The stages **after** glycolysis **differ**.

Glycolysis Makes Pyruvate from Glucose

1) Glycolysis involves splitting **one molecule** of glucose (with 6 carbons — 6C) into **two** smaller molecules of **pyruvate** (3C).
2) The process happens in the **cytoplasm** of cells.
3) Glycolysis is the **first stage** of both aerobic and anaerobic respiration and **doesn't need oxygen** to take place — so it's an **anaerobic** process.

There are Two Stages in Glycolysis — Phosphorylation and Oxidation

First, **ATP** is **used** to **phosphorylate glucose** to triose phosphate. Then **triose phosphate** is **oxidised, releasing ATP**. Overall there's a **net gain** of **2 ATP** and **2 reduced NAD**.

1) Stage One — Phosphorylation

1) Glucose is **phosphorylated** using a **phosphate** from a molecule of **ATP**. This creates **1** molecule of **glucose phosphate** and **1** molecule of **ADP**.
2) **ATP** is then used to add another **phosphate**, forming **hexose bisphosphate**.
3) **Hexose bisphosphate** is then **split** into 2 molecules of **triose phosphate**.

2) Stage Two — Oxidation

1) **Triose phosphate** is **oxidised** (loses hydrogen), forming 2 molecules of **pyruvate**.
2) **NAD** collects the hydrogen ions, forming **2 reduced NAD**.
3) **4 ATP** are **produced**, but 2 were used up in stage one, so there's a **net gain** of **2 ATP**.

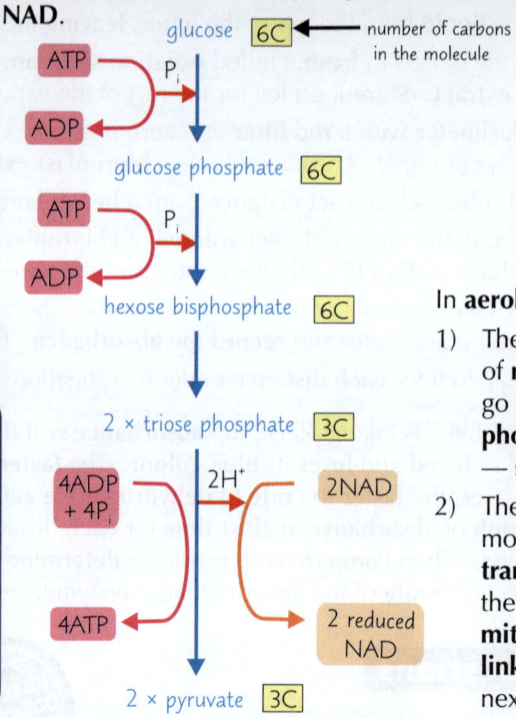

In **aerobic** respiration...

1) The **two** molecules of **reduced NAD** go to **oxidative phosphorylation** — see page 14.
2) The **two pyruvate** molecules are **actively transported** into the **matrix** of the **mitochondria** for the **link reaction** (see the next page).

In Anaerobic Respiration Pyruvate is Converted to Ethanol or Lactate

In **anaerobic** respiration, the **pyruvate** produced in glycolysis is **converted** into **ethanol** (in **plants** and **yeast**) or **lactate** (in **animal** cells and some **bacteria**) using **reduced NAD**:

Alcoholic Fermentation

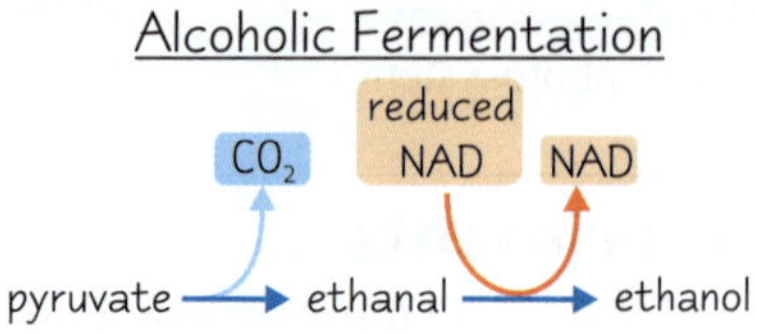

Lactate Fermentation

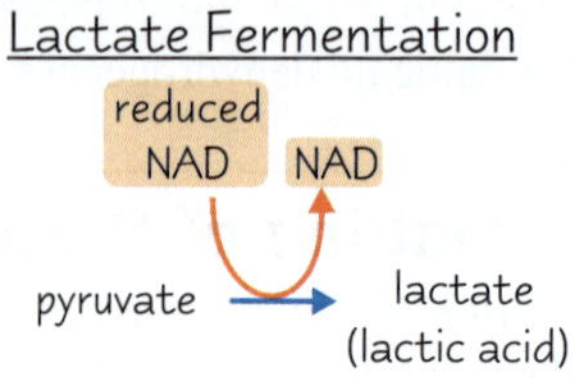

The production of ethanol or lactate **regenerates oxidised NAD**. This means **glycolysis** can **continue** even when there **isn't** much oxygen around, so a **small amount of ATP** can still be **produced** to keep some biological processes going... clever.

Aerobic Respiration

So, as you know from the previous page, in aerobic respiration, the two molecules of pyruvate from glycolysis enter the mitochondrial matrix for the link reaction. Here's what happens next...

The **Link Reaction** converts **Pyruvate** to **Acetyl Coenzyme A**

1) **Pyruvate** is **decarboxylated** (one carbon atom is **removed** from pyruvate in the form of CO_2).
2) **Pyruvate** is **oxidised** to form **acetate** and **NAD** is reduced to form **reduced NAD**.
3) **Acetate** is combined with **coenzyme A** (CoA) to form **acetyl coenzyme A** (**acetyl CoA**).
4) **No ATP** is produced in this reaction.

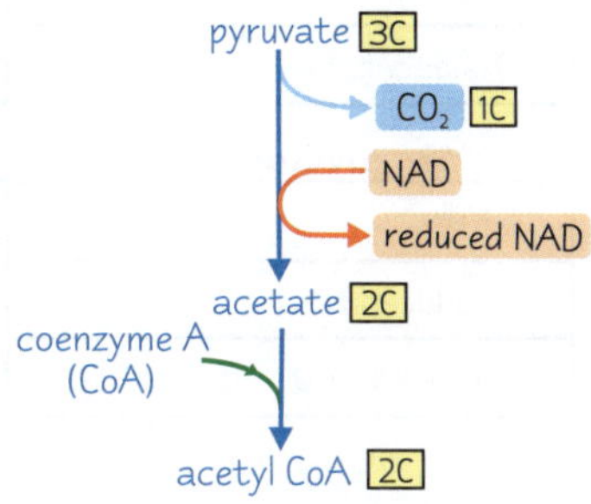

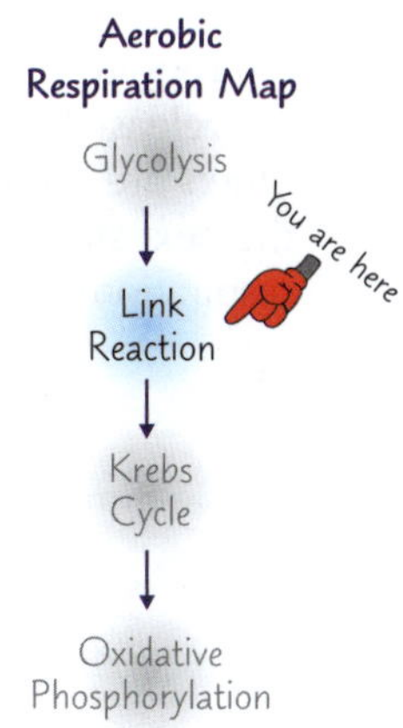

The **Link Reaction** Occurs **Twice** for Every **Glucose Molecule**

Two pyruvate molecules are made for **every glucose molecule** that enters glycolysis. This means the **link reaction** and the third stage (the **Krebs cycle**) happen **twice** for every glucose molecule. So for each glucose molecule:

- **Two** molecules of **acetyl coenzyme A** go into the Krebs cycle (see below).
- **Two CO_2 molecules** are released as a waste product of respiration.
- **Two** molecules of **reduced NAD** are formed and go to the last stage (oxidative phosphorylation, see page 14).

The **Krebs Cycle** Produces **Reduced Coenzymes** and **ATP**

The Krebs cycle involves a series of **oxidation-reduction reactions**, which take place in the **matrix** of the **mitochondria**. The cycle happens **once** for **every pyruvate** molecule, so it goes round **twice** for **every glucose** molecule.

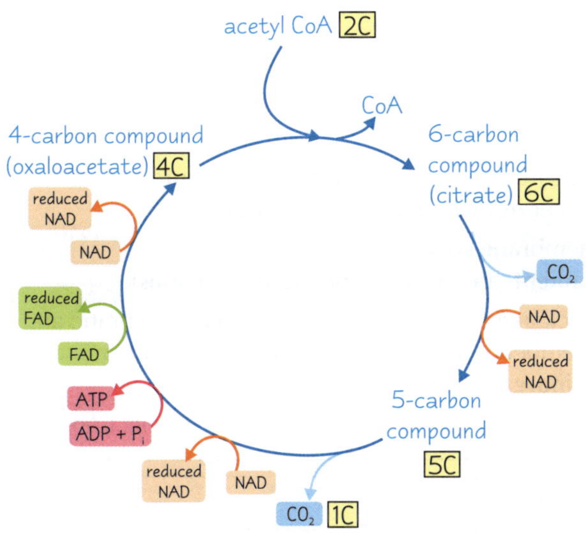

1
- **Acetyl CoA** from the link reaction combines with a **four-carbon molecule** (oxaloacetate) to form a **six-carbon molecule** (citrate).
- **Coenzyme A** goes back to the **link reaction** to be used again.

2
- The **6C citrate molecule** is converted to a **5C molecule**.
- **Decarboxylation** occurs, where CO_2 is **removed**.
- **Dehydrogenation** also occurs, where **hydrogen** is **removed**.
- The hydrogen is used to **produce reduced NAD** from NAD.

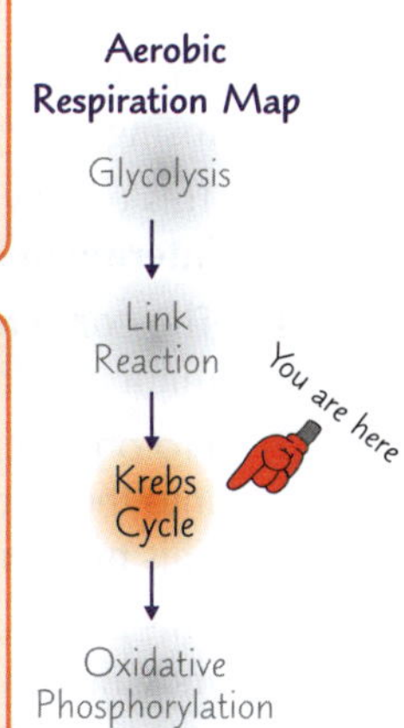

3
- The **5C molecule** is then converted to a **4C molecule**. (There are some intermediate compounds formed during this conversion, but you don't need to know about them.)
- **Decarboxylation** and **dehydrogenation** occur, producing **one** molecule of **reduced FAD** and **two** of **reduced NAD**.
- **ATP** is **produced** by the **direct transfer** of a **phosphate** group from an **intermediate** compound **to ADP**. When a phosphate group is directly transferred from one molecule to another it's called **substrate-level phosphorylation**. **Citrate** has now been **converted** into **oxaloacetate**.

Aerobic Respiration

Some **Products** of the **Krebs Cycle** are Used in **Oxidative Phosphorylation**

Some products are **reused**, some are **released** and others are used for the **next stage** of respiration:

Product from one Krebs cycle	Where it goes
1 coenzyme A	Reused in the next link reaction
Oxaloacetate	Regenerated for use in the next Krebs cycle
2 CO_2	Released as a waste product
1 ATP	Used for energy
3 reduced NAD	To oxidative phosphorylation
1 reduced FAD	To oxidative phosphorylation

Oxidative Phosphorylation Produces Lots of **ATP**

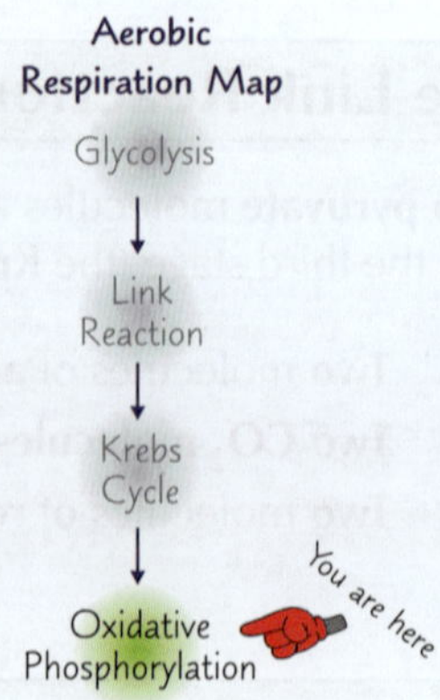

1) Oxidative phosphorylation is the process where the **energy** carried by **electrons**, from **reduced coenzymes** (reduced NAD and reduced FAD), is used to **make ATP**. (The whole point of the previous stages is to make reduced NAD and reduced FAD for the final stage).
2) Oxidative phosphorylation involves the **electron transport chain** and **chemiosmosis** (see below).

Protons are **Pumped Across** the **Inner Mitochondrial Membrane**

So now on to how **oxidative phosphorylation** actually **works**:

The regenerated coenzymes are reused in the Krebs cycle.

1) **Hydrogen atoms** are released from **reduced NAD** and **reduced FAD** as they're **oxidised** to NAD and FAD. The H atoms **split** into **protons (H^+)** and **electrons (e^-)**.
2) The **electrons** move down the **electron transport chain** (made up of **electron carriers**), **losing energy** at each carrier.
3) This energy is used by the electron carriers to **pump protons** from the **mitochondrial matrix into** the **intermembrane space** (the space **between** the inner and outer **mitochondrial membranes**).
4) The **concentration** of **protons** is now **higher** in the **intermembrane space** than in the mitochondrial matrix — this forms an **electrochemical gradient** (a **concentration gradient** of **ions**).
5) Protons then **move down** the **electrochemical gradient**, back across the inner mitochondrial membrane and into the mitochondrial matrix, via **ATP synthase** (which is embedded in the inner mitochondrial membrane). This **movement** drives the synthesis of **ATP** from **ADP** and **inorganic phosphate** (P_i).
6) This process of ATP production driven by the movement of H^+ ions across a membrane (due to electrons moving down an electron transport chain) is called **chemiosmosis** (which is described by the **chemiosmotic theory**).
7) In the mitochondrial matrix, at the end of the transport chain, the **protons**, **electrons** and **O_2** (from the blood) combine to form **water**. Oxygen is said to be the final **electron acceptor**.

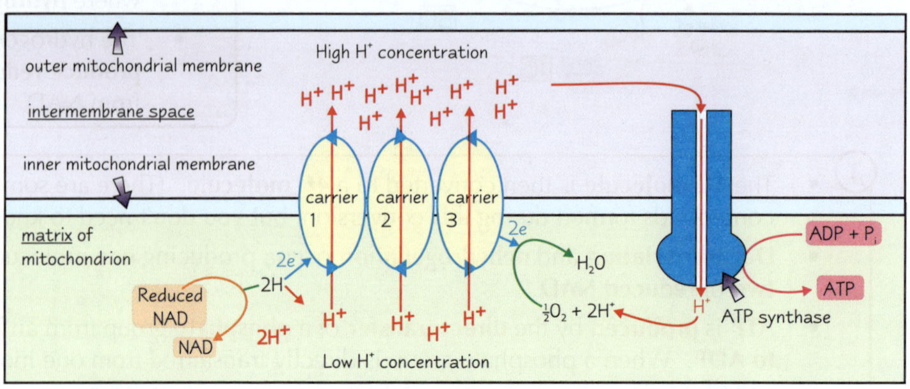

Aerobic Respiration

32 ATP Can be Made from One Glucose Molecule

As you know, **oxidative phosphorylation makes ATP** using energy from the reduced coenzymes — **2.5 ATP** are made from each **reduced NAD** and **1.5 ATP** are made from each **reduced FAD**. The table on the right shows **how much** ATP a cell can make from **one molecule** of **glucose** in **aerobic respiration**. (Remember, one molecule of glucose produces 2 pyruvate, so the link reaction and Krebs cycle happen twice.)

Stage of respiration	Molecules produced	Number of ATP molecules
Glycolysis	2 ATP	2
Glycolysis	2 reduced NAD	2 × 2.5 = 5
Link Reaction (×2)	2 reduced NAD	2 × 2.5 = 5
Krebs cycle (×2)	2 ATP	2
Krebs cycle (×2)	6 reduced NAD	6 × 2.5 = 15
Krebs cycle (×2)	2 reduced FAD	2 × 1.5 = 3
		Total ATP = 32

The number of ATP produced per reduced NAD or reduced FAD was thought to be 3 and 2, but new research has shown that the figures are nearer 2.5 and 1.5.

ATP Production Can be Affected by Mitochondrial Diseases

1) **Mitochondrial diseases** affect the **functioning** of **mitochondria**. They can affect how **proteins** involved in **oxidative phosphorylation** or the **Krebs cycle** function, **reducing ATP production**.
2) This may cause **anaerobic respiration** to **increase**, to try and make up some of the **ATP shortage**.
3) This results in **lots** of **lactate** being produced, which can cause **muscle fatigue** and **weakness**.
4) Some lactate will also **diffuse** into the **bloodstream**, leading to **high lactate concentrations** in the **blood**.

Other Respiratory Substrates Can also be Used in Aerobic Respiration

It's not just glucose that can be used as the **substrate** in aerobic respiration. Some products resulting from the **breakdown** of **other** molecules, such as **fatty acids** from **lipids** and **amino acids** from **proteins**, can be converted into molecules that are **able** to **enter** the **Kreb's cycle** (usually acetyl CoA).

Warm-Up Questions

PRACTICE QUESTIONS

Q1 Where in the cell does glycolysis occur?
Q2 Is glycolysis an anaerobic or aerobic process?
Q3 How many ATP molecules are used up in glycolysis?
Q4 What are the products of the link reaction?
Q5 Where in the cell does the Krebs cycle occur?
Q6 How many times does decarboxylation happen during one turn of the Krebs cycle?
Q7 What do the electrons lose as they move along the electron transport chain in oxidative phosphorylation?

Exam Questions

Q1 At the end of a 100 m sprint, runners will have built up lactate in their muscle cells.
a) Name the reduced coenzyme regenerated by lactate production. [1 mark]
b) What is the advantage for the runner of producing lactate in anaerobic respiration? [2 marks]

Q2 Carbon monoxide inhibits the final electron carrier in the electron transport chain.
a) Explain how this affects ATP production via the electron transport chain. [2 marks]
b) Explain how this affects ATP production via the Krebs cycle. [2 marks]

Q3 Describe how a 6-carbon molecule of glucose is converted to pyruvate. [6 marks]

The electron transport chain isn't just a FAD with the examiners...

Oh my gosh, I didn't think it could get any worse... You may be wondering how to learn these pages of crazy chemistry. Basically, you have to put in the time and go over and over it. Don't worry though, it WILL pay off and before you know it, you'll be set for the exams. And once you know this lot you'll be able to do anything, e.g. world domination.

PRACTICAL SKILLS

Respiration Experiments

You can use experiments to test how quickly respiration is taking place. Here are a few examples for you.

You can **Investigate Factors Affecting Respiration** in **Single-celled Organisms**

Yeast are single-celled organisms that can be grown in **culture**. They can respire **aerobically** when **plenty** of **oxygen** is **available** and **anaerobically** when **oxygen isn't available**. Both aerobic and anaerobic respiration in yeast produce **CO_2**, so the **rate of CO_2 production** gives an indication of the yeast's **respiration rate**. One way to measure CO_2 production is by using a **gas syringe** to collect the CO_2.

The methods below show you how to investigate the **effects** of **temperature** on **yeast respiration**. You'll need to decide what **temperatures** you're going to **investigate before** you **start** (e.g. 10 °C, 20 °C and 25 °C).

Aerobic Respiration

1) Put a **known volume** and **concentration** of **substrate solution** (e.g. glucose) in a test tube. Add a known volume of **buffer solution** to keep the **pH constant**. (Choose the optimum pH for the yeast you're testing — usually 4-6.)
2) Place the test tube in a **water bath** set to one of the **temperatures** being **investigated**. Leave it there for **10 minutes** to allow the temperature of the **substrate** to **stabilise**.
3) Add a **known mass** of **dried yeast** (e.g. *Saccharomyces cerevisiae*) to the test tube and **stir** for two minutes.
4) After the yeast has dissolved into the solution, put a **bung** with a **tube attached** to a **gas syringe** in the top of the test tube. The **gas syringe** should be set to **zero**.
5) **Start** a **stop watch** as soon as the bung has been put in the test tube.
6) As the yeast **respire**, the **CO_2 formed** will travel up the tube and into the **gas syringe**, which is used to measure the **volume** of **CO_2 released**.
7) At **regular time intervals** (e.g. every minute), record the **volume** of **CO_2** that is **present** in the **gas syringe**. Do this for a set amount of time (e.g. 10 minutes).
8) A **control** experiment should also be set up at each temperature, where **no yeast** is present. **No CO_2** should be formed without the yeast.
9) **Repeat** the experiment three times at **each temperature** you're investigating. Use your data to **calculate** the **mean rate of CO_2 production** at each temperature.

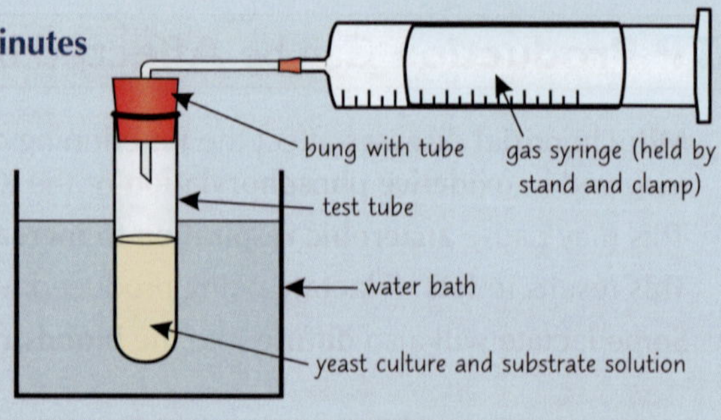

The yeast will only respire aerobically until the oxygen trapped in the tube is all used up. If you wanted to run the experiment for more time or with more yeast or glucose, you could use a conical flask that can trap more oxygen.

Anaerobic Respiration

1) Set up the apparatus according to **steps 1-3** of the experiment above.
2) After the yeast has dissolved into the substrate solution, trickle some **liquid paraffin** down the **inside** of the test tube so that it **settles** on and **completely covers** the **surface** of the solution. This will **stop oxygen** getting in, which will force the yeast to respire **anaerobically**.
3) Put a **bung**, with a **tube attached** to a **gas syringe**, in the top of the test tube. The **gas syringe** should be set to **zero**.
4) Perform **steps 5-9** from the method above.

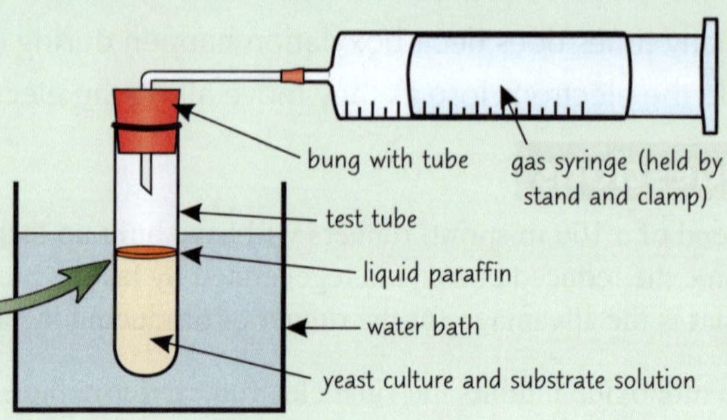

You can also easily **adapt** these methods to investigate the **effects** of other **variables**, such as **substrate concentration** and the use of **different respiratory substrates** (e.g. sucrose) on the **respiration rate**.

Just remember that you should **only change one variable at a time** (the independent variable, see p. 114). **All** the **other variables** that could affect your results need to be **controlled** (kept the same) or your results won't be **valid**.

Respiration Experiments

PRACTICAL SKILLS

The Rate of Oxygen Consumption can be Measured using a Respirometer

Respirometers can be used to indicate the **rate** of **aerobic respiration** by measuring the **amount** of **oxygen consumed** by an organism over a **period** of **time**. The example below shows how a respirometer can be used to measure the respiration rate of **woodlice**. You could also use it to measure the respiration rate of other small organisms or of plant seeds.

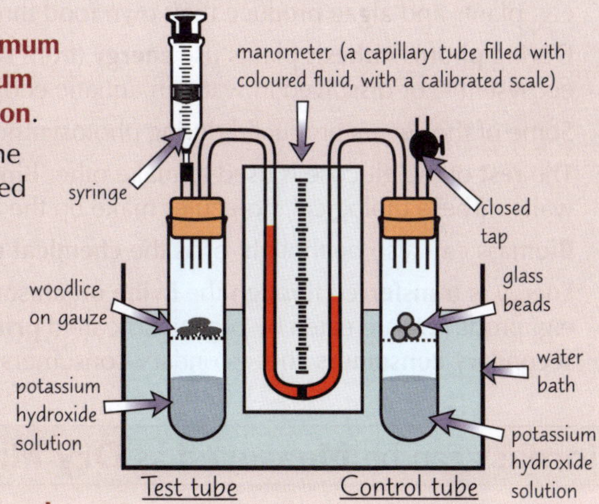

1) The apparatus is set up as shown on the right, partially submerged in a **water bath** at **15 °C** to provide the **optimum temperature** for the **woodlice** and therefore, the **optimum temperature** for the **enzymes** involved in their **respiration**.
2) The **control tube** is set up in exactly the **same way** as the woodlouse tube, except that the woodlice are substituted with **glass beads** of the **same mass**.
3) For **ten minutes**, the **tap** is **left open** and the **syringe** is **removed** to allow the **apparatus** to **equilibrate** (accounting for any **expansion** that might cause the **pressure** to **change inside**) and the **respiration rate** of the woodlice to **stabilise** in their new environment.
4) When the ten minutes is up, the **tap** is **closed** and the **syringe** is **attached**.
5) The **syringe** is used to **reset** the **manometer**, so that the **ends** of the **fluid** are at the **same level** on **either side** of the **'U'** and the reading from the **volume scale** on the **syringe** (usually in cm^3) is **recorded**.
6) As respiration occurs, the **volume** of the **air** in the test tube containing woodlice will decrease, due to the **oxygen consumed** during respiration (all the CO_2 produced is **absorbed** by the **potassium hydroxide**).
7) The decrease in the volume of the air will **reduce the pressure** in the test tube, causing the **coloured fluid** in the capillary tube of the manometer to **move towards** it.
8) After **leaving** the apparatus to **run** for a **set** period of time (e.g. 10 minutes), the syringe is used to **reset** the **manometer** and the **reading** on the **syringe's volume scale** is **recorded again**. The **difference** between **this figure** and the **figure taken** at the **start** of the **experiment** is the **oxygen consumption** for this **time period**. You can use this to calculate a **rate of respiration**.
9) To check the **precision** of the results, the experiment is **repeated** and a **mean volume** of O_2 is calculated.

Alfred the aphid thought holding his breath in the respirometer would be really funny. The students didn't.

Oxygen consumption can also be calculated by recording the movement of the fluid in the manometer, read from the scale on the manometer itself.

Warm-Up Questions

Q1 What does a respirometer measure?

Exam Question

PRACTICE QUESTIONS

Q1 A student was trying to find the optimum pH for yeast to produce ethanol. She set up three test tubes, each containing a solution of glucose buffered to a different pH. She then dissolved some dried *Saccharomyces cerevisiae* in the solution and trickled some liquid paraffin down the inside of the test tubes. Immediately after, she put a bung in the top of each test tube, with a tube attached to a gas syringe. Every 60 seconds, she recorded how much CO_2 had been released into the gas syringe.

a) Why did the student trickle liquid paraffin down the inside of the test tubes? [1 mark]

b) Why would measuring the rate of CO_2 production help her to find out how quickly ethanol was being produced? [2 marks]

c) Give two variables that should have been controlled in this experiment and describe how each of these variables should have been controlled. [2 marks]

d) What negative control should have been included in this experiment and why? [2 marks]

Respiration experiments — they're a gas...

Examiners love to ask you questions on experiments. Remember how these ones work in case something similar comes up in the exams. When you've got them stuck in your head, do something more interesting like learn to play the tuba.

Energy Transfer in Ecosystems

Some organisms get their energy from the Sun, some get it from other organisms, and it's all very friendly. Yeah right.

Plants **Photosynthesise** and Produce **Biomass**

1) An **ecosystem** includes all the **organisms** living in a particular area and all the **non-living** (abiotic) conditions (see p. 72).
2) In all ecosystems, there are **producers** — organisms that make their **own food**, e.g. plants and algae produce their own food through **photosynthesis**.
3) During **photosynthesis** plants use **energy** (from sunlight) and **carbon dioxide** (from the atmosphere in land-based ecosystems, or dissolved in water in aquatic ecosystems) to make **glucose** and other sugars (see pages 4-7).
4) Some of the sugars produced during photosynthesis are used in **respiration**, to release **energy** for **growth**.
5) The **rest** of the glucose is used to make other **biological molecules**, such as **cellulose** (a component of plant cell walls). These biological molecules make up the plant's **biomass** — the mass of living material.
6) **Biomass** can also be thought of as the **chemical energy stored** in the **plant**.
7) Energy is **transferred** through the **living organisms** of an ecosystem when organisms **eat** other organisms, e.g. producers are eaten by organisms called **primary consumers**. Primary consumers are then eaten by **secondary consumers** and secondary consumers are eaten by **tertiary consumers**. This is a **food chain** (see p. 20).

Biomass can be **Measured** as **Dry Mass** or Using a **Calorimeter**

Biomass can be measured in terms of the **mass of carbon** that an organism contains or the **dry mass** of its tissue per **unit area**.

> The water content of living tissue varies, so dry mass is used as a measure of biomass rather than wet mass.

1) **Dry mass** is the **mass** of the organism with the **water removed**.
2) To **measure** the dry mass, a **sample** of the organism is **dried**, often in an **oven** set to a low temperature. The sample is then weighed at **regular** intervals (e.g. every day). Once the **mass** becomes **constant** you know that all the water has been removed.
3) If needed, the result from the sample can be **scaled up** to give the dry mass (biomass) of the **total population** or the **area** being investigated. A **typical unit** for dry mass might be **kg m^{-2}**.
4) The **mass** of **carbon** present is generally taken to be **50%** of the dry mass.

You can **estimate** the amount of **chemical energy** stored in biomass by **burning** the **biomass** in a **calorimeter**. The amount of **heat given off** tells you **how much** energy is in it. Energy is measured in joules (J) or kilojoules (kJ).

1) A sample of dry biomass is **burnt** and the **energy released** is used to **heat** a **known volume** of **water**.
2) The **change in temperature** of the water is used to calculate the **chemical energy** of the **dry biomass**.

GPP and **NPP** are **Chemical Energy Stores**

> Remember plants convert light energy to chemical energy during photosynthesis.

1) **Gross primary production** (***GPP***) is the **total** amount of **chemical energy** converted from light energy by **plants**, in a given area.
2) Approximately 50% of the gross primary production is **lost to the environment** as **heat** when the plants **respire**. This is called **respiratory loss** (*R*).
3) The **remaining** chemical energy is called the **net primary production** (*NPP*). So ***NPP = GPP – R***.
4) The **NPP** is the energy available to the plant for **growth** and **reproduction** — the energy is stored in the plant's biomass. It is also the energy available to **organisms** at the **next stage** in the **food chain** (the next trophic level, see page 20). These include **herbivores** (animals that eat the plants) and **decomposers**.
5) Often primary production is expressed as a **rate** — i.e. the total amount of chemical energy (or biomass) in a given area, in a given **time**. Typical units might be **kJ ha^{-1} $year^{-1}$** (kilojoules per hectare per year) or **kJ m^{-2} yr^{-1}**. When primary production is expressed as a rate, it is called **primary productivity**.

net primary production = gross primary production – respiratory loss

EXAMPLE: The grass in an ecosystem has a gross primary productivity of **20 000 kJ m^{-2} yr^{-1}**. It loses **8 000 kJ m^{-2} yr^{-1}** as heat from **respiration**.

net primary productivity = 20 000 – 8 000
= 12 000 kJ m^{-2} yr^{-1}

Energy Transfer in Ecosystems

You Can Also Calculate Net Production for Consumers

Gus felt he needed to compensate for the 90% of energy he was not getting from his food.

1) **Consumers** also **store chemical energy** in their **biomass**.
2) Consumers get **energy** by **ingesting** plant material, or animals that have eaten plant material.
3) However, **not all** the chemical energy stored in the consumers' food is **transferred** to the **next** trophic level — around **90%** of the **total available energy** is **lost** in various ways.
4) Firstly, not all of the food is **eaten** (e.g. plant roots, bones) so the energy it contains is **not taken in**. Then, of the parts that **are ingested**:
   - Some are indigestible, so are **egested** as **faeces**. The **chemical energy** stored in these parts is therefore **lost** to the **environment**.
   - Some energy is also **lost** to the **environment** through **respiration** or excretion of **urine**.
5) The energy that's **left** after all this is **stored** in the **consumers' biomass** and is available to the next trophic level. This energy is the consumers' **net production**.
6) The **net production** of **consumers** can be **calculated** using the following **formula**:

Remember, when production is expressed as a rate it is called productivity.

$$N = I - (F + R)$$

N = Net production
I = Chemical energy in ingested food
F = Chemical energy lost in faeces and urine
R = Energy lost through respiration

EXAMPLE: The rabbits in an ecosystem ingest **20 000 kJ m^{-2} yr^{-1}** of energy, but lose **12 000 kJ m^{-2} yr^{-1}** of it in faeces and urine. They lose a further **6000 kJ m^{-2} yr^{-1}** using energy for **respiration**. You can use this to **calculate** the **net productivity** of the rabbits:

net productivity = 20 000 – (12 000 + 6000)
= 20 000 – 18 000
= 2000 kJ m^{-2} yr^{-1}

The net production of consumers can also be called **secondary production** (or **secondary productivity** when it's expressed as a **rate**).

7) You might also be asked to **calculate** how **efficient energy transfer** from one trophic level to another is:

The rabbits receive **20 000 kJ m^{-2} yr^{-1}**, and their **net productivity** is **2000 kJ m^{-2} yr^{-1}**. So the **percentage efficiency** of **energy transfer** is:

(2000 ÷ 20 000) × 100 = 10%

Warm-Up Questions

Q1 What is biomass?

Q2 How is energy transferred through an ecosystem?

Q3 State the formula for net primary production.

Q4 Briefly explain why not all the energy from one trophic level gets transferred to the next trophic level.

PRACTICE QUESTIONS

Exam Questions

Grass 13 883 kJ m^{-2}yr^{-1} → Arctic hare 2345 kJ m^{-2}yr^{-1} → Arctic fox 137 kJ m^{-2}yr^{-1}

Q1 The diagram above shows the net productivity of different trophic levels in a food chain.

a) Explain why the net productivity of the Arctic hare is less than the net primary productivity of the grass. [4 marks]

b) The Arctic hare ingests 18 905 kJ m^{-2} yr^{-1} of food. Calculate the total energy loss of the Arctic hare. [2 marks]

Q2 A farmer grows cabbages in one of his fields.

a) Suggest how he could estimate the chemical energy store in the dry mass of one of his cabbages. [3 marks]

b) Using this estimate, the energy of the cabbage field was calculated as 15 600 kJ m^{-2}. Does this represent the gross or net primary production? Give a reason for your answer. [2 marks]

Boy, do I need an energy transfer this morning...

Golly, lots of similar sounding terms on these pages. Plants are primary producers, so you calculate their net primary production. They get energy from the Sun and lose some through respiration. For consumers it's just net production — they eat, then lose energy from respiration and faeces and urine. Simple. I mean, you never saw a plant on the loo...

Farming Practices and Production

Farmers know the theory behind energy transfers and try to use it to maximise production — smart thinking. You don't have to milk the cows, but you do need to know how to increase the efficiency of energy transfer...

Food Webs Show How Energy is Transferred Between Organisms

1) **Food chains** and **food webs** show how energy is **transferred** through an ecosystem.
2) **Food chains** show **simple lines** of energy transfer. Each of the stages in a food chain is called a **trophic level**.
3) **Food webs** show **lots of food chains** in an ecosystem and how they **overlap**.
4) **Decomposers** (e.g. fungi) are also part of food webs. Decomposers **break down dead** or **undigested** material, allowing nutrients to be recycled (see page 22).

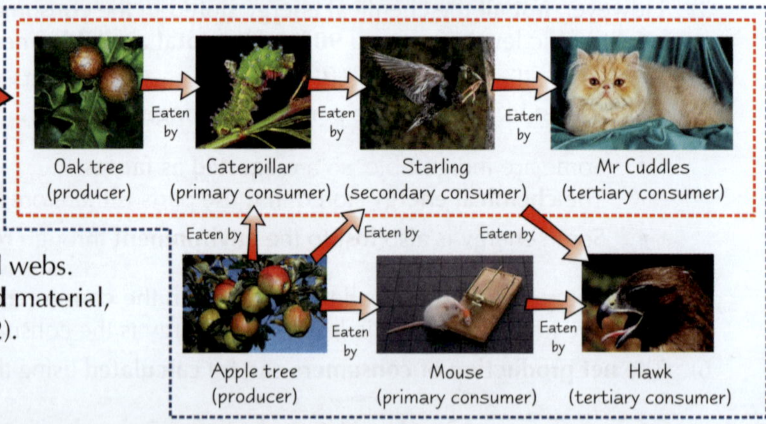

Farming Practices Increase The Efficiency of Energy Transfer

Most farming practices aim to **increase** the amount of **energy** that is **available** for **human consumption**. There are different ways this can be done. You need to know about **two** of them.

1) The **energy lost** to other **organisms**, e.g. pests, can be **reduced**.
2) The **energy lost** through **respiration** can be **reduced**.

1 Simplifying Food Webs Reduces Energy Loss to Other Organisms

Here's an example of a simplified **food web** involving a **crop plant** grown for **human consumption**:

The weed, the mouse and the aphid are **pests** — organisms that **reduce** the amount of **energy** available for **crop growth** and therefore the **net primary production** (NPP) of crops. This ultimately **reduces** the amount of **energy** available for **humans**. By **simplifying** the food web, i.e. getting rid of food chains that **don't** involve humans, **energy losses** will be **reduced** and the **NPP** of the crop will **increase**.

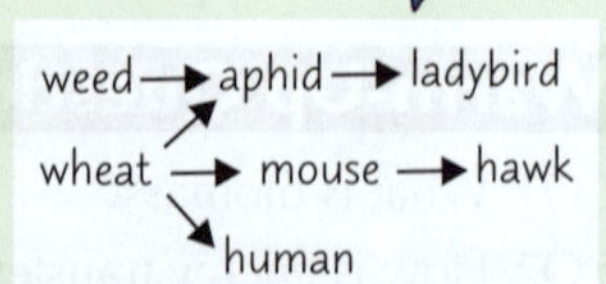

Simplifying the food web means **getting rid of pests** — and for that, farmers need **pest control**.

1) Farmers can **reduce pest numbers** using **chemical pesticides**. For example:

- **Insecticides** kill **insect** pests that **eat** and **damage** crops. Killing insect pests means **less biomass** is **lost** from crops, so they grow to be **larger**, which means **NPP** is **greater**.
- **Herbicides** kill **weeds** (unwanted plant species). Killing weeds can **remove direct competition** with the crop for energy from the Sun. It can also remove the preferred habitat or food source of the **insect** pests, helping to **further reduce** their numbers and **simplify** the food web.

2) **Biological agents also** reduce the **numbers of pests**, so crops lose **less energy** and **biomass**, **increasing** the efficiency of energy transfer to humans.

- **Parasites** live in or lay their **eggs** on a **pest insect**. Parasites either **kill** the insect or **reduce** its ability to **function**, e.g. some wasp species lay their eggs inside caterpillars — the eggs hatch and **kill** the caterpillars.
- **Pathogenic** (disease-causing) **bacteria** and **viruses** are used to kill pests, e.g. the bacterium ***Bacillus thuringiensis*** produces a **toxin** that kills a wide range of **caterpillars**.

Natural predators can also be introduced to the ecosystem to eat the pest species, e.g. ladybirds eat aphids — this is useful but doesn't really simplify the food web.

3) Farmers can use **integrated systems** that combine **both** chemical and biological methods. The **combined effect** of using both can reduce pest numbers **even more** than either method **alone**, meaning **NPP** is **increased** even more.

Farming Practices and Production

2 Reducing Respiratory Losses Means Energy is Transferred More Efficiently

1) One way that farmers **increase** the **net production** of their **livestock** is by **controlling** the **conditions** that they live in, so that **more** of their **energy** is used for **growth** and **less** is **lost** through **respiration** (and activities that **increase** the **rate** of respiration). For example:

- Movement increases the rate of respiration, so animals may be kept in pens where their **movement** is **restricted**.
- The pens are often **indoors** and **kept warm**, so **less energy** is **wasted by** generating body heat.

Increasing production was not an issue that was easy to raise with Herbert.

2) This means that **more biomass** is produced and **more chemical energy** can be stored, **increasing net production** and the **efficiency** of **energy transfer** to humans.

The benefits are that **more food** can be produced in a **shorter** space of time, often at **lower cost**. However, enhancing net production by keeping animals in pens raises **ethical issues**. For example, some people think that the **conditions** intensively reared animals are kept in cause the animals **pain**, **distress** or restricts their **natural behaviour**, so it **shouldn't be done**.

Warm-Up Questions

Q1 What is a food web?

Q2 What is the role of decomposers in a food web?

Q3 How does simplifying a food web involving a crop increase the NPP of the crop?

PRACTICE QUESTIONS

Exam Question

Q1 The graph below shows the yearly percentage loss of a crop to three different insect pests before and after a chemical pesticide was used on the crop.

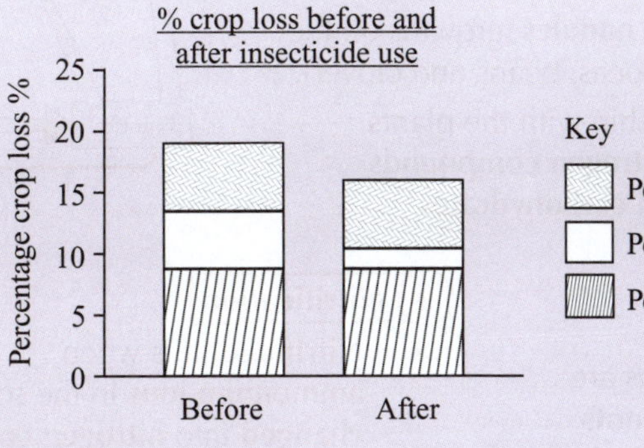

a) How do insect pests reduce the NPP of crops? [1 mark]

b) What conclusions can be drawn about the effectiveness of the chemical pesticide from this graph? [2 marks]

c) Suggest two ways in which the farmer growing this crop could further reduce the percentage crop loss to insect pests. [2 marks]

d) Explain two ways in which livestock farmers can increase the net production of their animals. [2 marks]

Farming practices — baa-aa-aa-rmy...

Crikey, so farming's not just about getting up early to feed the chicks then — farmers want to produce as much food as they can, so they try to eliminate energy losses to pests and respiration. Remember, farmers really want to maximise production — the more energy available for crop and livestock growth, the better.

Nutrient Cycles

Organisms don't need to worry about which recycling bin to use. Ecosystems have developed a much better system to make sure necessary elements like nitrogen and phosphorus can be recycled and don't run out.

Fungi and **Bacteria** Have an **Important Role** in **Nutrient Recycling**

1) A **natural ecosystem** is one that hasn't been **changed** by **human activity**. In **natural ecosystems** nutrients are recycled though the food webs, but **human activity** often **disrupts** the cycling of nutrients.
2) **Microorganisms**, such as **bacteria** and **fungi**, are an important part of food webs. Many are **saprobionts** (a type of decomposer) — they feed on the **remains** of **dead plants** and **animals** and on their **waste products** (faeces, urine), breaking them down. This allows important **chemical elements** in the remains to be **recycled**.
3) Saprobionts **secrete enzymes** and **digest** their **food externally**, then **absorb** the **nutrients** they need. This is known as **extracellular digestion**. During this process, **organic molecules** are broken down into **inorganic ions**. Obtaining nutrients from dead organic matter using extracellular digestion is known as **saprobiotic nutrition**.
4) Some fungi form **symbiotic relationships** with the **roots** of **plants**. These relationships are known as **mycorrhizae**.
   - The fungi are made up of **long, thin strands** called **hyphae**, which **connect** to the **plant's roots**.
   - The hyphae greatly **increase** the **surface area** of the plant's root system, helping the plant to absorb **ions** from the soil that are usually **scarce** (e.g. **phosphorus**). Hyphae also **increase** the uptake of **water** by the plant.
   - In turn, the fungi obtain **organic compounds**, such as glucose, from the plant.

The **Nitrogen Cycle** shows how **Nitrogen** is **Recycled** in **Ecosystems**

Plants and animals **need nitrogen** to make **proteins** and **nucleic acids** (DNA and RNA). The atmosphere's made up of about 78% nitrogen gas, but plants and animals **can't use it** in that form — they need **bacteria to convert** it into **nitrogen-containing compounds** first. The **nitrogen cycle** shows how nitrogen is **converted** into a usable form and then **passed** on between different **living** organisms and the **non-living** environment.

The nitrogen cycle includes **food chains** (nitrogen is passed on when organisms are eaten), and four different processes that involve bacteria — **nitrogen fixation**, **ammonification**, **nitrification** and **denitrification**:

1) Nitrogen fixation

- **Nitrogen fixation** is when nitrogen **gas** in the atmosphere is turned into nitrogen-containing compounds. Biological nitrogen fixation is carried out by **bacteria** such as ***Rhizobium***. They turn nitrogen into **ammonia**, which goes on to form ammonium ions in solution that can then be **used** by plants.
- *Rhizobium* are found inside **root nodules** (growths on the roots) of **leguminous** plants (e.g. peas, beans and clover).
- They form a **mutualistic** relationship with the plants — they provide the plant with **nitrogen compounds** and the plant provides them with **carbohydrates**.

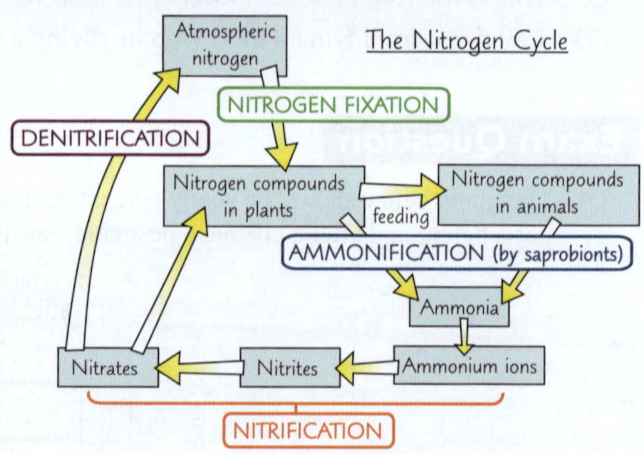

2) Ammonification

- **Ammonification** is when nitrogen compounds from **dead organisms** are turned into **ammonia** by **saprobionts**, which goes on to form **ammonium ions**.
- Animal **waste** (**urine** and **faeces**) also contains nitrogen compounds. These are also turned into ammonia by saprobionts and go on to form ammonium ions.

Don't worry — you don't need to learn the names of the microorganisms.

3) Nitrification

- **Nitrification** is when **ammonium ions** in the soil are **changed** into **nitrogen compounds** that can then be **used** by plants (nitrates).
- First **nitrifying bacteria** called ***Nitrosomonas*** change **ammonium ions** into **nitrites**.
- Then other nitrifying bacteria called ***Nitrobacter*** change **nitrites** into **nitrates**.

4) Denitrification

- **Denitrification** is when nitrates in the soil are **converted** into **nitrogen gas** by **denitrifying bacteria** — they use nitrates in the soil to carry out **respiration** and produce nitrogen gas.
- This happens under **anaerobic conditions** (where there's **no** oxygen), e.g. in **waterlogged** soils.

Other ways that **nitrogen** gets into an **ecosystem** are by **lightning** (which **fixes atmospheric nitrogen**) or by **artificial fertilisers** (they're **produced from atmospheric nitrogen** on an **industrial scale** in the **Haber process**).

Nutrient Cycles

Phosphorus is Passed Through the **Food Web** in The **Phosphorus Cycle**

Plants and **animals** need **phosphorus** to make **biological molecules** such as **phospholipids** (which make up cell membranes), **DNA** and **ATP**. Phosphorus is found in **rocks** and **dissolved** in the **oceans** in the form of **phosphate ions** (PO_4^{3-}). Phosphate ions dissolved in water in the soil can be **assimilated** (absorbed and then used to make more complex molecules) by **plants** and other **producers**.

The phosphorus cycle shows how phosphorus is passed through an **ecosystem**.

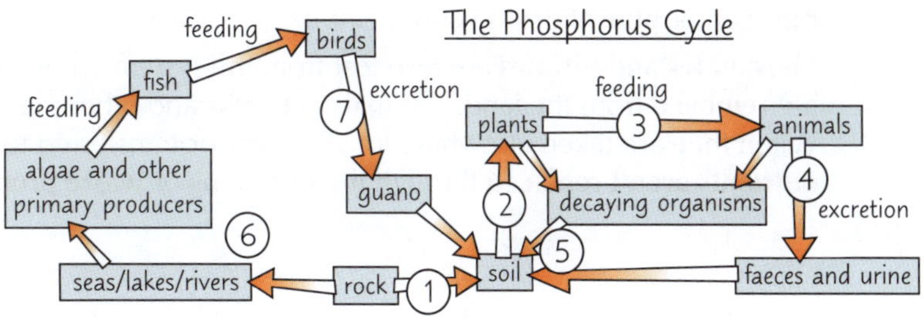

1) Phosphate ions in **rocks** are released into the soil by **weathering**.
2) Phosphate ions are **taken into** the plants through the **roots**. **Mycorrhizae** (see previous page) greatly **increase** the **rate** at which phosphorus can be assimilated.
3) Phosphate ions are **transferred** through the **food chain** as animals eat the plants and are in turn eaten by other animals.
4) Phosphate ions are **lost** from the **animals** in waste **products**.
5) When plants and animals **die**, **saprobionts** are involved in **breaking down** the organic compounds, releasing **phosphate ions** into the soil for **assimilation** by **plants**. These microorganisms also release the phosphate ions from **urine** and **faeces**.
6) **Weathering** of **rocks** also releases phosphate ions into **seas**, **lakes** and **rivers**. This is taken up by **aquatic producers**, such as algae, and passed along the food chain to **birds**.
7) The waste produced by **sea birds** is known as **guano** and contains a **high proportion** of phosphate ions. Guano returns a significant amount of phosphate ions to **soils** (particularly in coastal areas). It is often used as a **natural fertiliser**.

Warm-Up Questions

Q1 What are saprobionts?

Q2 How do mycorrhizae benefit plants?

Q3 Why do plants and animals need nitrogen?

Q4 Briefly describe the process of nitrification.

Q5 How do animals obtain phosphate ions?

Q6 How are phosphate ions transferred from the sea to the land?

PRACTICE QUESTIONS

Exam Question

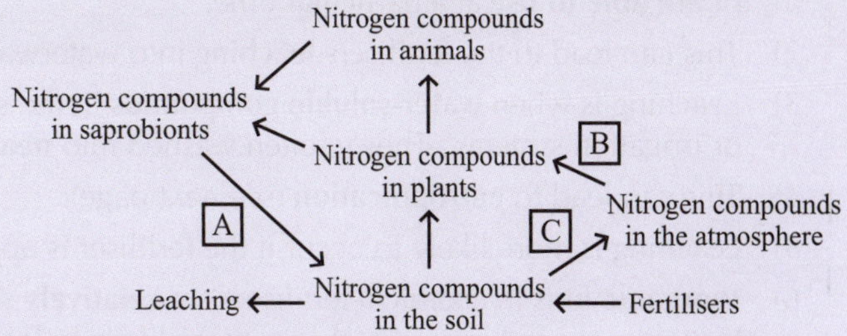

Q1 The diagram on the right shows the nitrogen cycle.

a) Name the processes labelled A, B and C in the diagram. [3 marks]

b) i) Describe the role of saprobionts in process A. [2 marks]

ii) Describe how saprobionts obtain their nutrients. [2 marks]

Nitrogen fixation — cheaper than a shoe fixation...

The nitrogen cycle's not as bad as it seems. Divide up the four processes of nitrogen fixation, ammonification, nitrification and denitrification and learn them separately, then hey presto — you've learnt the whole cycle. Learning the phosphorus cycle ain't that bad either — it's got a few rocks in it as well as all the plants and animals though.

Fertilisers and Eutrophication

Every silver lining has a dark cloud — using fertilisers to replace lost nutrients is all fine and dandy till they don't stay where you put 'em and end up killing all the fish...

Nutrients are Lost when Crops are Harvested

1) Crops **take in** minerals from the soil as they **grow** and use them to build their own tissues.
2) When **crops** are **harvested**, they're **removed** from the field where they're grown rather than being allowed to die and decompose there. This means the **mineral ions** that they contain (e.g. **phosphates** and **nitrates**) are not returned to the **soil** by **decomposers** in the **nitrogen** or **phosphorus cycles**.
3) **Phosphates** and **nitrates** are also lost from the system when **animals** or **animal products** are **removed** from the land. Animals eat **grass** and other plants, **taking in** their **nutrients**. When they are taken elsewhere for slaughter or transferred to a different field, the **nutrients aren't replaced** through their **remains** or **waste products**.

Fertilisers are Added to Soils to Replace Lost Nutrients

Adding fertiliser **replaces** the lost minerals, so **more energy** from the ecosystem can be used for growth, **increasing** the **efficiency** of energy transfer. Fertilisers can be **artificial** or **natural**.

Artificial fertilisers are **inorganic** — they contain **pure chemicals** (e.g. ammonium nitrate) as powders or pellets.

Natural fertilisers are **organic** matter — they include **manure**, **composted vegetables**, **crop residues** (the parts left over after the harvest) and **sewage sludge**.

True, Daisy had a dippy fringe, but she sure could produce a lot of fertiliser.

Using Fertilisers Raises Environmental Issues

1) Sometimes **more** fertiliser is **applied** than the plants **need** or are **able to use** at a particular time.
2) This can lead to the fertilisers **leaching** into waterways.
3) Leaching is when **water-soluble compounds** in the **soil** are washed away, e.g. by rain or irrigation systems. They're often washed into **nearby ponds** and **rivers**.
4) This can lead to **eutrophication** (see next page).
5) Leaching is more likely to occur if the fertiliser is applied **just before heavy rainfall**.
6) **Inorganic ions** in chemical fertilisers are **relatively soluble**. This means that excess minerals that are **not used immediately** are **more likely** to leach into waterways. In **natural fertilisers**, the **nitrogen** and **phosphorus** are still contained in **organic molecules** that need to be **decomposed** by microorganisms before they can be absorbed by plants. This means that their **release** into the soil for uptake by plants is more **controlled**, and leaching is **less likely**.
7) The **leaching** of **phosphates** is **less likely** than the leaching of **nitrates** because phosphates are **less soluble** in water.
8) Using fertilisers also changes the **balance** of **nutrients** in the soil — **too much** of a particular nutrient can cause crops and other plants to **die**.

Fertilisers and Eutrophication

Eutrophication is Caused by Excess Nutrients

This is the process of **eutrophication**:

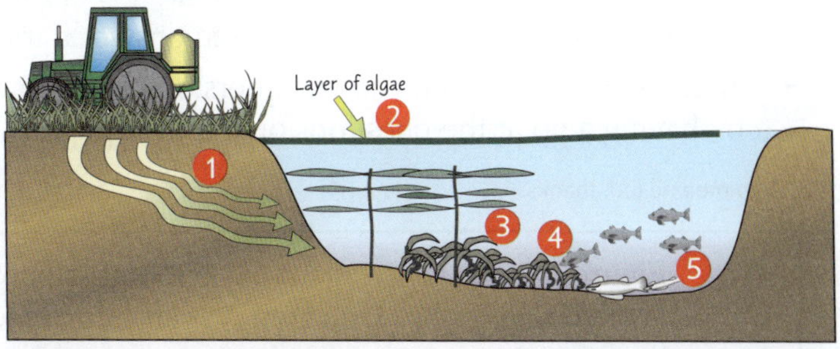

1. **Mineral ions leached** from **fertilised fields** stimulate the **rapid growth** of **algae** in ponds and rivers.
2. Large amounts of algae **block light** from reaching the plants below.
3. Eventually the **plants die** because they're **unable** to **photosynthesise** enough.
4. **Bacteria** feed on the dead plant matter. The **increased** numbers of **bacteria reduce** the **oxygen** concentration in the water by carrying out **aerobic respiration**.
5. **Fish** and other aquatic organisms **die** because there **isn't enough dissolved oxygen**.

Hey, who turned out the lights?

Warm-Up Questions

Q1 Why are nutrients lost when plants are harvested?

Q2 What are artificial fertilisers?

Q3 What is leaching?

Q4 Briefly describe the process of eutrophication.

Exam Question

Q1 A study was conducted to investigate the effect, on a nearby river, of adding fertiliser to farmland. The oxygen and algal content of a river that runs past a field where nitrate fertiliser had been applied was measured at the field and up to a distance of 180 m away. A similar control river next to an unfertilised field was also studied. The results are shown in the graphs below.

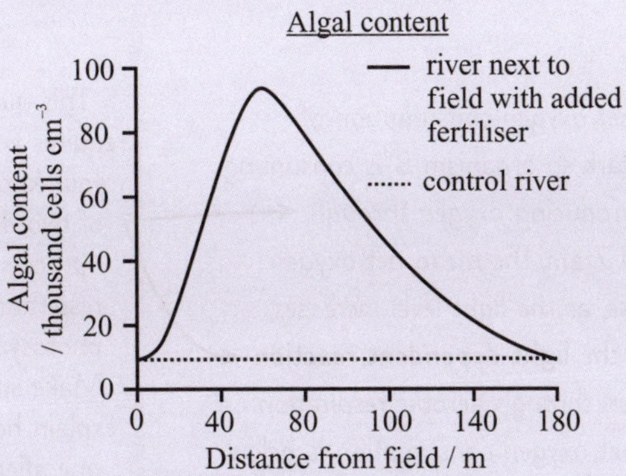

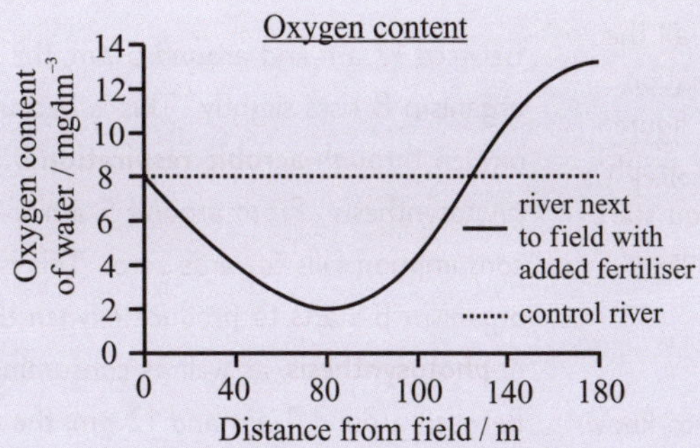

a) Explain the purpose of the control river in the study. [1 mark]

b) Calculate the percentage increase in algal content from 0 to 60 m away from the fertilised field. [1 mark]

c) Describe the relationship between the algal content of the water and the oxygen content of the water in the river next to the fertilised field. [1 mark]

d) Suggest an explanation for the relationship you described in part c). [4 marks]

Help — everything I just learnt is leaching out of my brain...

Fertilisers are important for giving plants all the nutrients they need but, as with a lot of things, a little goes a long way. Too much fertiliser and you can find yourself struggling to breathe. Literally. If you're a fish that is. Nitrogen and phosphorus are good for algae as well as plants and if they get hold of it, by 'eck do their numbers explode...

Extra Exam Practice

Phew, Topic 5 was a tricky one — there're lots of complex processes to remember, that's for sure. I find the best way to check I've understood everything is to just dive right in to some practice exam questions. Ok, maybe re-energise with a quick cuppa and a biscuit* first — I'll wait...

- Have a look at this example of how to answer a tricky exam question.
- Then check how much you've understood from Topic 5 by having a go at the questions on the next page.

*Grab me one too, thanks.

When you're feeling ready, there are synoptic questions on pages 123-127 that will test your knowledge of the whole course.

1 **Figure 1** shows the mean net oxygen consumption for two organisms (**A** and **B**), over 24 hours, in the UK in spring. Organism **A** is an animal species and organism **B** is a plant species.

Figure 1

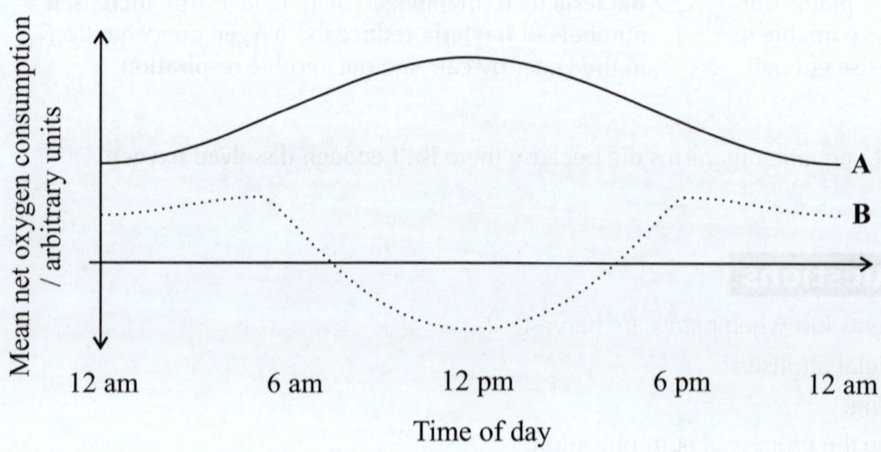

1.1 Explain what the graph shows for organism **B**, between the hours of 12 am and 12 pm.

(6 marks)

1.2 Explain how the rate of acetyl coenzyme A production in organism **A** will change between 12 pm and 6 pm.

(2 marks)

Take your time to read all the information and to figure out the graph before you start your answer.

1.1

Between 12 am and around 5 am, the mean net oxygen consumption of organism B rises slightly. This is because it's dark so organism B is consuming oxygen through **aerobic respiration** but not producing oxygen through photosynthesis. From around 5 am to around 7 am, the mean net oxygen consumption falls towards zero. This is because, as the light level increases, organism B starts to produce oxygen through the **light-dependent reaction** of **photosynthesis**, as well as consuming oxygen through aerobic respiration. Between around 7 am and 12 pm, the mean net oxygen consumption is below zero, meaning that organism B is producing more oxygen through photosynthesis than it is using up through aerobic respiration.

This question requires you to use your knowledge of two different processes — respiration and photosynthesis. Make sure you explain how each one affects the graph in your answer.

You need to know the names of lots of different substances as part of your course. If you get asked about a specific substance in a question, take time to think about its role and how it relates to what you're being asked.

1.2

Acetyl coenzyme A is produced during the **link reaction** of **aerobic respiration**. As the mean net oxygen consumption in organism A falls between 12 pm and 6 pm, this suggests that the rate of aerobic respiration is falling and therefore that the rate of the link reaction and acetyl coenzyme A production is falling too.

Extra Exam Practice

2 Aquaponics is a farming technique in which fish and plants are grown together in a controlled environment. This technique reduces the need for artificial fertilisers, as waste products from the fish can supply the plants with essential nutrients. **Figure 2** shows the basic set-up of an aquaponics system.

Figure 2

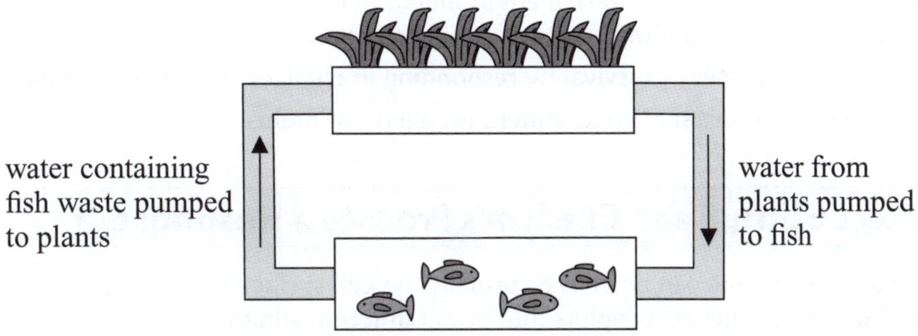

In an aquaponics system, constant artificial lighting can be used to aid photosynthesis.

2.1 Describe how light energy is used to produce ATP during photosynthesis. Details of electron transfer chains are not required.

(2 marks)

2.2 Explain how constant artificial lighting could increase the production of organic substances in the plants.

(4 marks)

Microorganisms are an essential component of an aquaponics system.

2.3 Explain why saprobionts and nitrifying bacteria are needed to ensure the plants are able to synthesise molecules containing nitrogen and phosphorus.

(5 marks)

3 A simplified food chain is shown in **Figure 3**.

Figure 3

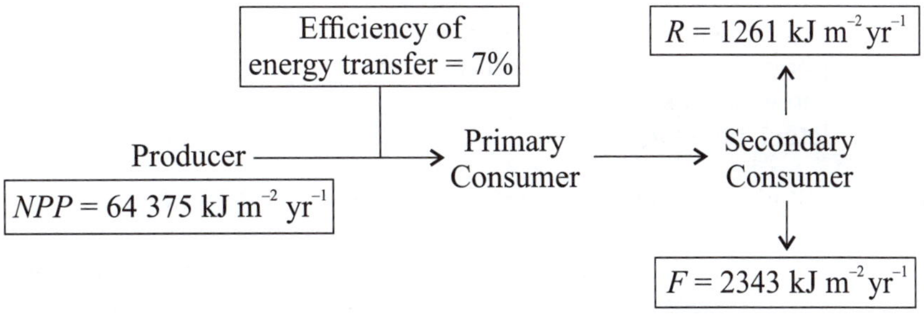

3.1 During summer, the *NPP* of the producer increases. This is partly due to an increased activity rate of rubisco. Explain how an increase in the activity rate of rubisco could lead to a higher *NPP*.

(3 marks)

3.2 The secondary consumer does not ingest 10% of the material available in the primary consumer. Using this information and **Figure 3**, calculate the net productivity of the secondary consumer.

(3 marks)

3.3 The breathing rate of the secondary consumer increases when it is hunting for food. Explain how this helps to maximise the rate of ATP synthesis in the secondary consumer.

(3 marks)

Nervous Communication

Your body has an amazing network of nerve cells which constantly send electrical signals — a bit like a big circuit board.

Responding to their Environment Helps Organisms Survive

1) **Animals increase** their **chances** of **survival** by **responding** to **changes** in their **external environment**, e.g. by **avoiding harmful environments** such as places that are too hot or too cold.
2) They also **respond** to **changes** in their **internal environment** to make sure that the **conditions** are always **optimal** for their **metabolism** (all the chemical reactions that go on inside them).
3) **Plants** also **increase** their **chances** of **survival** by **responding** to **changes** in their **environment** (see p. 30).
4) Any **change** in the internal or external **environment** is called a **stimulus**.

Receptors Detect Stimuli and Effectors Produce a Response

1) **Receptors detect stimuli** — they can be **cells** or **proteins** on **cell surface membranes**. There are **loads** of **different types** of receptors that detect **different stimuli**.
2) **Effectors** are cells that bring about a **response** to a **stimulus**, to produce an **effect**. Effectors include **muscle cells** and cells found in **glands**, e.g. the **pancreas**.
3) Receptors **communicate** with effectors via the **nervous system** or the **hormonal system**, or sometimes using **both**.

Receptors are specific to one type of stimulus — see p. 32.

The Nervous System Sends Information as Electrical Impulses

1) The **nervous system** is made up of a **complex network** of cells called **neurones**. There are **three main types:**
   - **Sensory neurones** transmit electrical impulses from **receptors** to the **central nervous system (CNS)** — the **brain** and **spinal cord**.
   - **Motor neurones** transmit electrical impulses from the **CNS** to **effectors**.
   - **Relay neurones** transmit electrical impulses **between** sensory neurones and motor neurones.

Electrical impulses are also called nerve impulses or action potentials (see p. 36).

2) A stimulus is detected by **receptor cells** and an **electrical impulse** is sent along a **sensory neurone**.
3) When an **electrical impulse** reaches the end of a neurone, chemicals called **neurotransmitters** take the information across to the **next neurone**, which then sends an **electrical impulse** (see p. 39).
4) The **CNS** (the **coordinator**) **processes** the information and sends impulses along **motor neurones** to an **effector**.

Stimulus → **Receptors** → (sensory neurone) → **CNS** → (motor neurone) → **Effectors** → **Response**

- **Stimulus**: E.g. you see a friend **waving**.
- **Receptors**: **Light receptors** (photoreceptors) in your eyes **detect** the wave.
- **CNS**: CNS **processes information**.
- **Effectors**: **Muscle cells** are stimulated by the motor neurones.
- **Response**: **Muscles contract** to make your arm **wave**.

5) The **nervous system** is split into two different systems:

You don't need to learn the structure of the nervous system, but understanding it'll help with the rest of the section.

- The **central nervous system (CNS)** — made up of the **brain** and the **spinal cord**.
- The **peripheral nervous system** — made up of the neurones that connect the CNS to the **rest** of the **body**. It also has two different systems:
  - The **somatic nervous system** controls **conscious** activities, e.g. running and playing video games.
  - The **autonomic nervous system** controls **unconscious** activities, e.g. digestion. It's got two divisions that have **opposite effects** on the body:
    - The **sympathetic** nervous system gets the body **ready for action**. It's the '**flight or fight**' system.
    - The **parasympathetic** nervous system **calms** the body down. It's the '**rest and digest**' system.

Harold thought it was about time his sympathetic nervous system took over.

Nervous Communication

Reflexes are Rapid, Automatic Responses to Stimuli

1) A **reflex** is where the body **responds** to a stimulus **without** making a **conscious decision** to respond.
2) Because you don't have to **spend time deciding** how to respond, information travels **really fast** from **receptors** to **effectors**.
3) So simple reflexes help organisms to **protect** the body because they're **rapid**.
4) The **pathway** of neurones linking receptors to effectors in a reflex is called a **reflex arc**. You need to **learn** a **simple reflex arc** involving three neurones — a **sensory**, a **relay** and a **motor** neurone.

E.g. the hand-withdrawal response to heat

- **Thermoreceptors** in the skin **detect** the **heat** stimulus.
- The **sensory neurone** carries impulses to the **relay neurone**.
- The **relay neurone** connects to the **motor neurone**.
- The **motor neurone** sends **impulses** to the **effector** (your biceps muscle).
- Your **muscle contracts** to withdraw your hand and **stop** it being **damaged**.

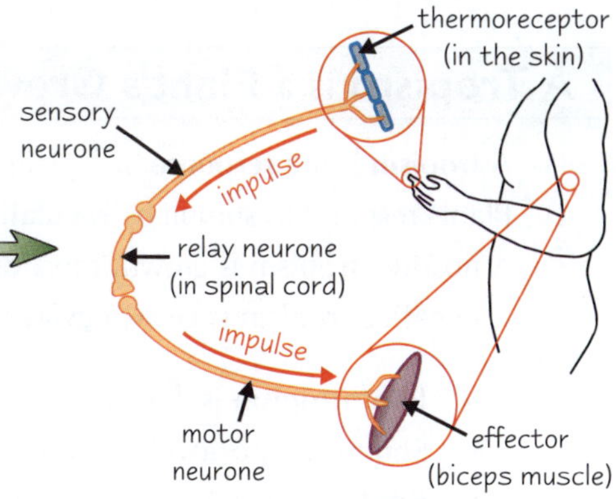

5) If there's a **relay neurone** involved in the simple reflex arc then it's possible to **override** the reflex, e.g. in the example above your **brain** could tell your hand to **withstand** the **heat**.

Nervous System Communication is Localised, Short-lived and Rapid

1) When an **electrical impulse** reaches the end of a neurone, **neurotransmitters** (see page 39) are **secreted directly** onto **target cells** (e.g. muscle cells) — so the nervous response is **localised**.
2) **Neurotransmitters** are **quickly removed** once they've done their job, so the response is **short-lived**.
3) Electrical impulses are **really fast**, so the response is **rapid** — this allows animals to **react quickly** to stimuli.

Warm-Up Questions

Q1 What is a stimulus?

Q2 Name the three main types of neurone.

Q3 What is a reflex?

Q4 How can reflexes help protect the body?

Exam Questions

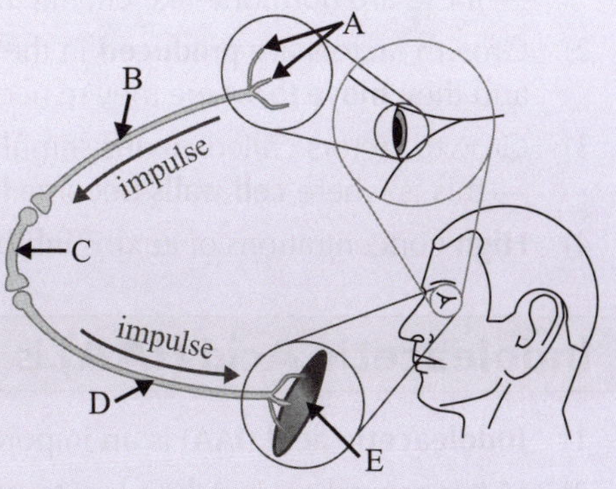

Q1 An animal responds to a stimulus in its environment.
State the role of receptors and effectors in this response. [2 marks]

Q2 The human blink reflex is an involuntary response, which results in the automatic closing of the eyelids (a blink) when an object touches the surface of the eye. A reflex arc for the blink reflex is shown in the diagram above.

a) Using the diagram, describe the reflex arc involved in this response. [4 marks]

b) The knee-jerk is another reflex response. You can test for it by tapping someone just below their patella (knee cap). Suggest why the absence of this response could indicate some damage to a person's CNS. [1 mark]

Q3 Polio is a virus that can cause damage to the CNS. In severe cases, the virus can damage motor neurones. Suggest and explain how this might lead to paralysis. [3 marks]

Responding to questions in an exam helps you to pass...

Actually, this stuff is really quite fascinating once you realise just how much your body can do without you even knowing. Just sit back and let your nerves do the work... Ah, apart from the whole revision thing — your body can't do that without you knowing, unfortunately. Get your head around these pages before you tackle the rest of the section.

Responses in Plants and Animals

Plants and simple animals respond to uncomplicated things like gravity and light. This helps them to survive in their environment...

Plants Need to **Respond** to **Stimuli Too**

Flowering plants, like animals, **increase** their chances of **survival** by **responding** to changes in their **environment**, e.g:

- They sense the direction of **light** and **grow** towards it to **maximise** light absorption for **photosynthesis**.
- They can sense **gravity**, so their roots and shoots **grow** in the **right direction**.
- **Climbing** plants have a sense of **touch**, so they can find things to climb up and **reach** the **sunlight**.

A **Tropism** is a **Plant's Growth Response** to an **External Stimulus**

1) A **tropism** is the **response** of a plant to a **directional stimulus** (a stimulus coming from a particular direction).
2) Plants respond to stimuli by **regulating** their **growth**.
3) A **positive tropism** is growth **towards** the stimulus.
4) A **negative tropism** is growth **away** from the stimulus.

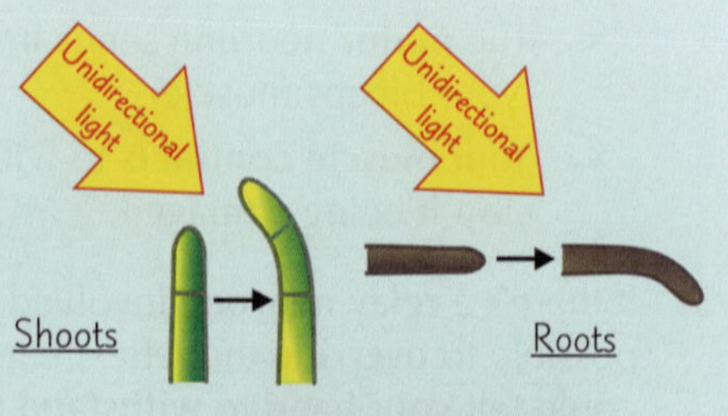

- **Phototropism** is the growth of a plant in response to **light**.
- **Shoots** are **positively phototropic** and grow **towards** light.
- **Roots** are **negatively phototropic** and grow **away** from light.

- **Gravitropism** is the growth of a plant in response to **gravity**.
- **Shoots** are **negatively gravitropic** and grow **upwards**.
- **Roots** are **positively gravitropic** and grow **downwards**.

The men's gymnastics team were negatively gravitropic.

Responses are **Brought About** by **Growth Factors**

1) Plants **respond** to directional stimuli using specific **growth factors** — these are hormone-like chemicals that **speed up** or **slow down** plant **growth**.
2) Growth factors are **produced** in the **growing regions** of the plant (e.g. shoot tips, leaves) and they **move** to where they're needed in the **other parts** of the plant.
3) Growth factors called **auxins** stimulate the **growth** of shoots by **cell elongation** — this is where **cell walls** become **loose** and **stretchy**, so the cells get **longer**.
4) **High** concentrations of auxins **inhibit growth** in **roots** though.

Indoleacetic Acid (IAA) is an Important **Auxin**

1) **Indoleacetic acid** (**IAA**) is an important **auxin** that's produced in the **tips** of **shoots** in flowering plants.
2) IAA is **moved** around the plant to **control tropisms** — it moves by **diffusion** and **active transport** over short distances, and via the **phloem** over long distances.
3) This results in **different parts** of the plant having **different concentrations** of IAA. The **uneven distribution** of IAA means there's **uneven growth** of the plant, e.g:

Phototropism — IAA moves to the more **shaded** parts of the **shoots** and **roots**, so there's uneven growth.

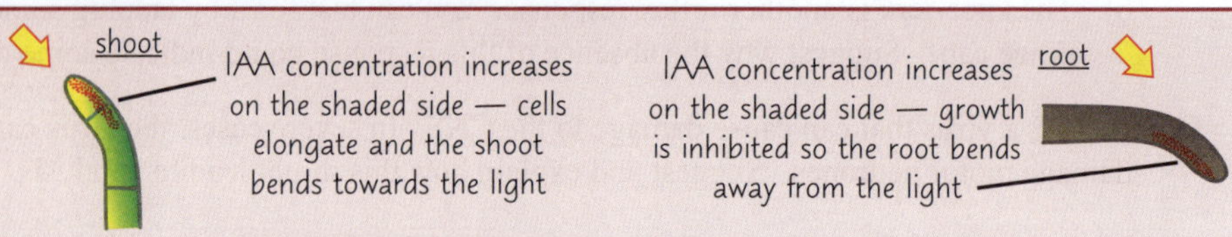

Gravitropism — IAA moves to the **underside** of **shoots** and **roots**, so there's uneven growth.

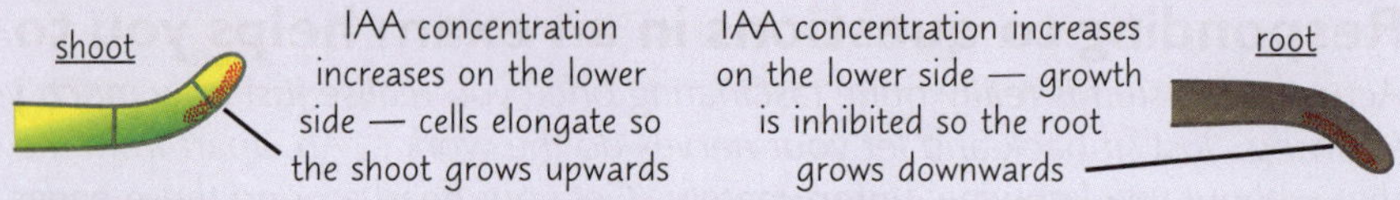

Responses in Plants and Animals

Simple Responses Keep Simple Organisms in a Favourable Environment

Simple mobile organisms, e.g. woodlice and earthworms, have **simple responses** to keep them in a **favourable environment**. Their **response** can either be **tactic** or **kinetic**:

- **Tactic responses** (**taxes**) — the organisms move towards or away from a **directional stimulus**, e.g. **light**.

  For example, **woodlice** show a **tactic** response to light (**phototaxis**) — they move **away from** a **light source**. This helps them **survive** as it keeps them **concealed** under stones during the day (where they're **safe** from predators) and keeps them in **damp conditions** (which reduces water loss).

- **Kinetic responses** (**kineses**) — the organisms' movement is affected by a **non-directional** stimulus, e.g. **humidity**.

  For example, **woodlice** show a **kinetic** response to **humidity**. In **high humidity** they move **slowly** and **turn less** often, so that they **stay where they are**. As the air gets **drier**, they move **faster** and **turn more** often, so that they move into a **new area**. This response **increases** the **chance** that a woodlouse will move to an area with **higher humidity**. This **improves** the **survival** chances of the organism — it **reduces** their **water loss** and it helps to keep them **concealed**.

You Can Use Choice Chambers to Investigate Animal Responses

A **choice chamber** is a container with different compartments, in which you can create different **environmental conditions**. It can be used to investigate how animals, such as woodlice, **respond** to conditions like **light intensity** or **humidity** in the **laboratory**. Here's how you can use a choice chamber:

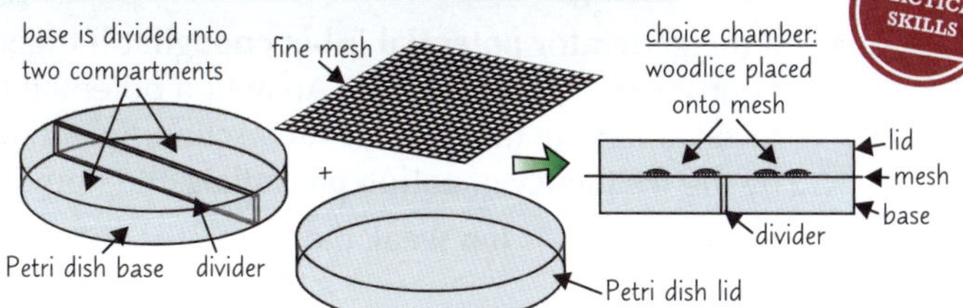

1) Construct a choice chamber using the **equipment** shown in the diagram.
2) To investigate the effect of **light intensity** on woodlouse movement, cover **one half** of the lid (including the sides) with **black paper**. This will make **one side** of the chamber **dark**. Put **damp filter paper** in **both** sides of the base.
3) Place **10 woodlice** on the mesh in the centre of the chamber and cover the chamber with the lid.
4) After **10 minutes**, take off the lid and record the number of woodlice on each side of the chamber.
   Try to minimise the amount of time the lid is off, so that the environmental conditions created aren't disturbed.
5) **Repeat** the experiment after gently moving the woodlice back to the centre. You should find that most woodlice **end up** on the **dark side** of the choice chamber (a **tactic response** to light).
   You can use a small, soft paintbrush to help with moving the woodlice if necessary. For ethical reasons, you should handle the woodlice carefully and return them to their natural habitat as soon as possible. Make sure you wash your hands after handling the woodlice.
6) To investigate **humidity**, place some **damp filter paper** in one side of the base and a **desiccating** (**drying**) **agent** in the other side. **Don't cover** the **lid** with paper. Put the **lid on** and leave the chamber for 10 minutes to stabilise before carrying out steps 3)-5) above.
7) You can do a similar experiment using a **maze** instead of a choice chamber.

Warm-Up Questions

Q1 What is positive gravitropism?

Q2 Describe the difference between taxes and kineses.

Exam Question

Week	Height of plant not given auxins / cm	Height of plant provided with auxins / cm
1	1	2
2	2	5
3	4	8
4	6	9
5	9	13

Q1 The table shows the results some students obtained when they investigated the effect of providing plants with auxins.

a) Describe and explain what the data shows. [2 marks]

b) Suggest why this data might be useful to a commercial tomato producer. [1 mark]

c) Explain the role of auxins in the control of phototropism in the shoots. [3 marks]

IAA Productions — do you have the growth factor — with Simon Trowel...

The tactic response to revision — when you see your revision notes, you always move away from them. Or if you were a plant, I guess we could say you were negatively revisi-tropic. You've still got to learn this lot for your exams though.

Receptors

Receptors are the front line in animal responses — they detect what's going on and pass on information about it.

Receptors are Specific to One Kind of Stimulus

1) Receptors are **specific** — they only **detect one particular stimulus**, e.g. light, pressure or glucose concentration.
2) There are **many different types** of receptor that each detect a **different type of stimulus.**
3) Some receptors are **cells**, e.g. photoreceptors are receptor cells that connect to the nervous system. Some receptors are **proteins** on **cell surface membranes**, e.g. glucose receptors are proteins found in the cell membranes of some pancreatic cells.
4) Here's a bit more about how receptor cells that communicate information via the **nervous system** work:
   - When a nervous system receptor is in its **resting state** (not being stimulated), there's a **difference in charge** between the **inside** and the **outside** of the cell — this is generated by ion pumps and ion channels (see p. 36). This means that there's a **voltage** across the membrane. Voltage is also known as **potential difference.**
   - The **potential difference** when a cell is at **rest** is called its **resting potential.** When a stimulus is detected, the cell membrane is **excited** and becomes **more permeable**, allowing **more ions** to move **in** and **out** of the cell — **altering** the **potential difference**. The **change** in **potential difference** due to a stimulus is called the **generator potential.**
   - A **bigger stimulus** excites the membrane more, causing a **bigger movement** of ions and a **bigger change** in potential difference — so a **bigger generator potential** is produced.
   - If the **generator potential** is **big enough** it'll trigger an **action potential** — an electrical impulse along a neurone (see pages 36-37). An action potential is only triggered if the generator potential reaches a certain level called the **threshold** level. Action potentials are all one size, so the **strength** of the **stimulus** is measured by the **frequency** of **action potentials** (the number of action potentials triggered during a certain time period).
   - If the stimulus is **too weak** the generator potential **won't reach** the **threshold**, so there's **no action potential.**

Pacinian Corpuscles are Pressure Receptors in Your Skin

1) **Pacinian corpuscles** are **mechanoreceptors** — they detect **mechanical stimuli**, e.g. **pressure** and **vibrations.** They're found in your **skin.**
2) Pacinian corpuscles contain the end of a **sensory neurone**, imaginatively called a **sensory nerve ending.** The sensory nerve ending is **wrapped** in loads of layers of connective tissue called **lamellae.**
3) When a Pacinian corpuscle is **stimulated**, e.g. by a tap on the arm, the lamellae are **deformed** and **press** on the **sensory nerve ending.**
4) This causes the sensory neurone's cell membrane to stretch, deforming the **stretch-mediated sodium ion channels.** The channels **open** and **sodium ions diffuse into** the cell, creating a **generator potential.**
5) If the **generator potential** reaches the **threshold**, it triggers an **action potential.**

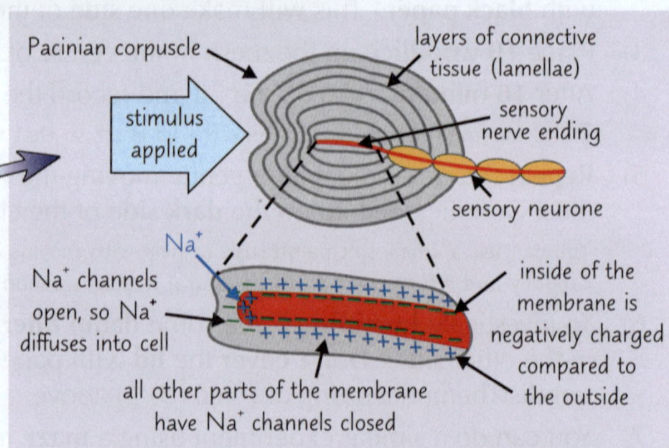

Pacinian corpuscles only respond to mechanical stimuli — this is a good example of how receptors only respond to specific stimuli.

Photoreceptors are Light Receptors in Your Eye

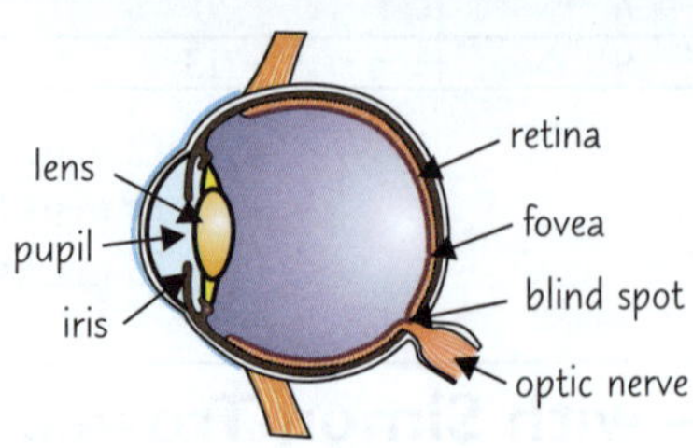

1) **Light** enters the eye through the **pupil.** The **amount** of light that enters is **controlled** by the muscles of the **iris.**
2) Light rays are **focused** by the **lens** onto the **retina**, which lines the inside of the eye. The retina contains **photoreceptor cells** — these **detect light.**
3) The **fovea** is an area of the retina where there are **lots** of **photoreceptors.**
4) **Nerve impulses** from the photoreceptor cells are carried from the **retina** to the **brain** by the **optic nerve**, which is a bundle of **neurones.** Where the optic nerve leaves the eye is called the **blind spot** — there **aren't** any **photoreceptor cells**, so it's **not sensitive** to **light.**

Receptors

Photoreceptors Convert **Light** into an **Electrical Impulse**

1) **Light** enters the eye, hits the **photoreceptors** and is **absorbed** by **light-sensitive optical pigments**.
2) Light bleaches the pigments, causing a **chemical change** and altering the **membrane permeability** to **sodium** ions.
3) A **generator potential** is created and if it reaches the threshold, a nerve impulse is sent along a **bipolar neurone**.
4) Bipolar neurones connect **photoreceptors** to the **optic nerve**, which takes impulses to the **brain**.

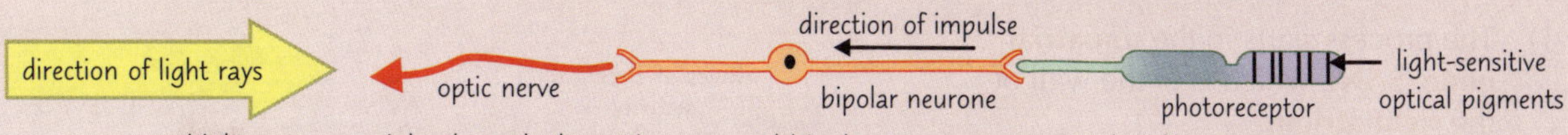

Light passes straight through the optic nerve and bipolar neurone to get to the photoreceptor.

5) The human eye has **two types** of photoreceptor — **rods** and **cones**.
6) Rods are mainly found in the **peripheral** parts of the **retina**, and cones are found **packed together** in the **fovea**.
7) Rods and cones contain **different optical pigments** making them **sensitive** to **different wavelengths** of light.
8) Rods only give information in **black and white** (monochromatic vision), but cones give information in **colour** (trichromatic vision). There are three types of cones, each containing a different **optical pigment** — **red-sensitive**, **green-sensitive** and **blue-sensitive**. When they're stimulated in **different proportions** you see different colours.

Rods are **More Sensitive**, but **Cones** let you See **More Detail**

Sensitivity

- Rods are **very sensitive to light** (they work well in **dim light**). This is because **many rods** join **one neurone**, so many weak **generator potentials combine** to **reach** the **threshold** and trigger an action potential.

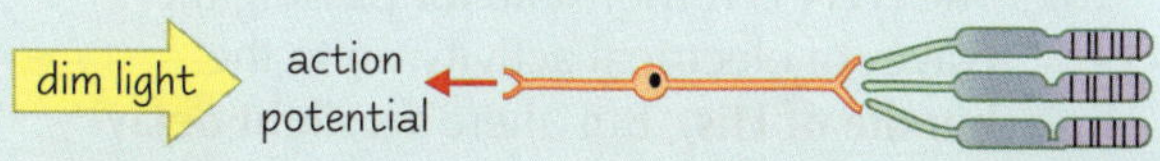

- Cones are **less sensitive** than rods (they work best in **bright light**). This is because **one cone** joins **one neurone**, so it takes more light to reach the threshold and trigger an action potential.

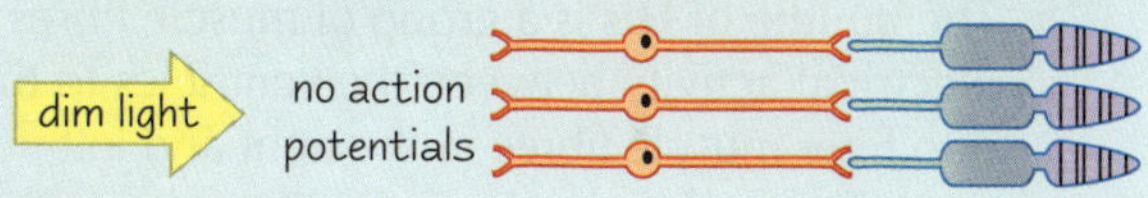

Visual acuity (the ability to tell apart points that are close together)

- Rods give **low visual acuity** because **many rods** join the **same neurone**, which means light from two points close together **can't** be told apart.

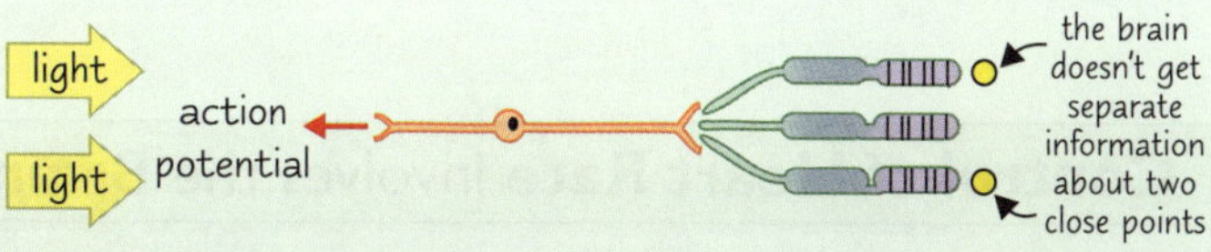

- Cones give **high visual acuity** because cones are **close together** and **one cone** joins **one neurone**. When light from two points hits two cones, two action potentials (one from **each cone**) go to the brain — so you can distinguish two points that are close together as **two separate points**.

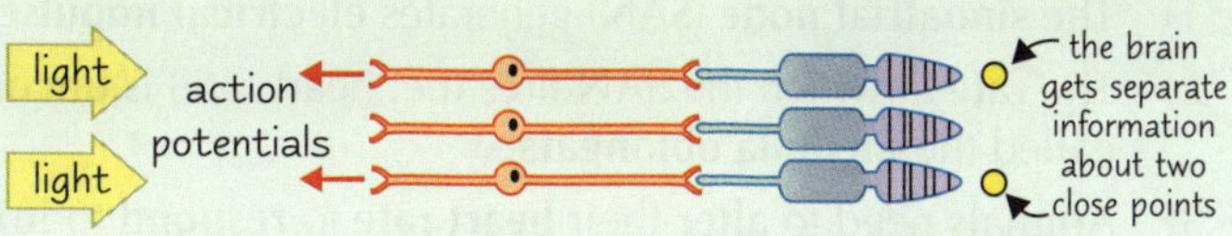

Warm-Up Questions

Q1 Why are receptors described as specific?

Q2 In the human eye, which type of photoreceptor is more sensitive to light?

PRACTICE QUESTIONS

Exam Questions

Q1 Explain how a generator potential is created when a Pacinian corpuscle is stimulated. [3 marks]

Q2 Explain how the human eye can provide high visual acuity. [3 marks]

Pacinian corpuscles love deadlines — they work best under pressure...

Wow, loads of stuff here, so cone-gratulations if you manage to remember it all. Receptors are really important because without them you wouldn't be able to see this book, and without this book revision would be way trickier.

Control of Heart Rate

You don't have to think about making your heart beat — thankfully your heart does it by itself. However, your body has systems to control your heart beat which speed it up or slow it down.

Cardiac Muscle Controls the **Regular Beating** of the Heart

Cardiac (heart) muscle is '**myogenic**' — it can contract and relax without receiving signals from nerves. This pattern of contractions controls the **regular heartbeat**.

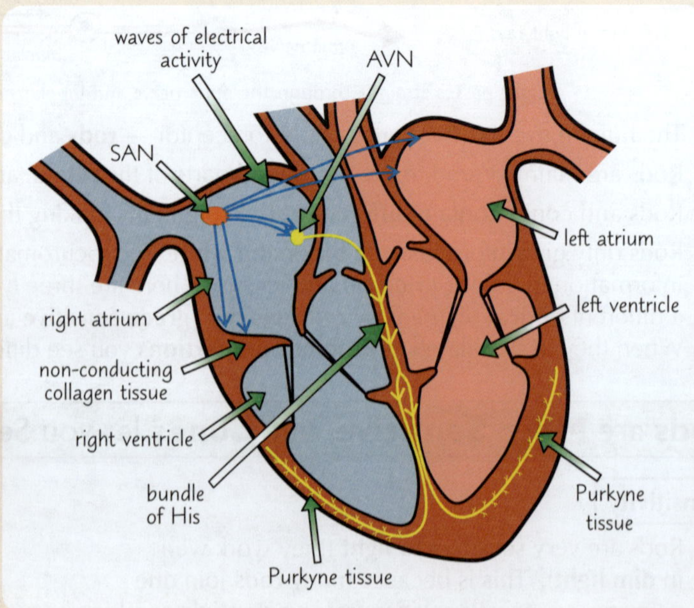

1) The process starts in the **sinoatrial node** (**SAN**), which is in the wall of the **right atrium**.
2) The SAN is like a pacemaker — it sets the **rhythm** of the heartbeat by sending out regular **waves of electrical activity** to the atrial walls.
3) This causes the right and left **atria** to **contract at the same time**.
4) A band of non-conducting **collagen tissue** prevents the waves of electrical activity from being passed directly from the atria to the ventricles.
5) Instead, these waves of electrical activity are transferred from the SAN to the **atrioventricular node** (**AVN**).
6) The AVN is responsible for passing the waves of electrical activity on to the **bundle of His**. But, there's a **slight delay** before the AVN reacts, to make sure the atria have emptied **before** the ventricles contract.
7) The **bundle of His** is a group of muscle fibres responsible for conducting the waves of electrical activity between the ventricles to the apex (bottom) of the heart. The bundle splits into finer muscle fibres in the right and left ventricle walls, called the **Purkyne tissue**.
8) The Purkyne tissue carries the waves of electrical activity into the muscular walls of the right and left ventricles, causing them to **contract simultaneously**, from the bottom up.

Control of Heart Rate Involves the **Brain** and **Autonomic Nervous System**

1) The **sinoatrial node** (**SAN**) generates **electrical impulses** that cause the **cardiac muscles** to **contract**.
2) The **rate** at which the SAN fires (i.e. heart rate) is **unconsciously controlled** by a part of the **brain** called the **medulla oblongata**.
3) Animals need to **alter** their **heart rate** to **respond** to **internal stimuli**, e.g. to prevent fainting due to low blood pressure or to make sure the heart rate is high enough to supply the body with enough oxygen.
4) **Stimuli** are **detected** by **pressure receptors** and **chemical receptors**:
   - There are **pressure receptors** called **baroreceptors** in the **aorta** and the **carotid arteries** (major arteries in the neck). They're stimulated by **high** and **low blood pressure**.
   - There are **chemical receptors** called **chemoreceptors** in the **aorta**, the **carotid arteries** and in the **medulla**. They **monitor** the **oxygen** level in the **blood** and also **carbon dioxide** and **pH** (which are indicators of O_2 level).

There's more about the autonomic nervous system on page 28.

5) Electrical impulses from receptors are sent **to the medulla** along **sensory** neurones. The medulla processes the information and sends impulses to the SAN along **sympathetic** or **parasympathetic** neurones (which are part of the **autonomic nervous system**). There's more on this on the next page.

Control of Heart Rate

Stimuli Detected by Receptors Cause Heart Rate to Speed Up or Slow Down

This table shows how the heart responds to different **stimuli**:

Stimulus	Receptor	Neurone and transmitter	Effector	Response
High blood pressure.	**Baroreceptors** detect **high** blood pressure.	Impulses are sent to the medulla, which sends impulses along **parasympathetic** neurones. These secrete **acetylcholine** (a neurotransmitter), which binds to receptors on the SAN.	Cardiac muscles	Heart rate **slows** down to **reduce blood pressure** back to normal.
Low blood pressure.	**Baroreceptors** detect **low** blood pressure.	Impulses are sent to the medulla, which sends impulses along **sympathetic** neurones. These secrete **noradrenaline** (a neurotransmitter), which binds to receptors on the SAN.	Cardiac muscles	Heart rate **speeds up** to **increase blood pressure** back to normal.
High blood O_2, **low** CO_2 or **high** pH levels.	**Chemoreceptors** detect chemical changes in the blood.	Impulses are sent to the medulla, which sends impulses along **parasympathetic** neurones. These secrete **acetylcholine**, which binds to receptors on the SAN.	Cardiac muscles	Heart rate **decreases** to return O_2, CO_2 and pH levels back to normal.
Low blood O_2, **high** CO_2 or **low** pH levels.	**Chemoreceptors** detect chemical changes in the blood.	Impulses are sent to the medulla, which sends impulses along **sympathetic** neurones. These secrete **noradrenaline**, which binds to receptors on the SAN.	Cardiac muscles	Heart rate **increases** to return O_2, CO_2 and pH levels back to normal.

For more about neurotransmitters see pages 39-41.

When Ed did that special thing to her beak, Polly's sympathetic neurones went into overdrive.

Warm-Up Questions

Q1 Why is heart muscle described as 'myogenic'?

Q2 What is the function of the bundle of His?

Q3 Why do animals need to alter their heart rate?

Q4 Name the effectors that are involved in increasing or decreasing heart rate.

Exam Questions

PRACTICE QUESTIONS

Q1 The control of heart rate is coordinated by specific parts of the heart. Describe the function of:

a) the sinoatrial node. [1 mark]

b) the Purkyne tissue. [1 mark]

Q2 Exercise causes an increase in the levels of carbon dioxide in the blood.

a) Explain how increased blood CO_2 leads to an increased heart rate. [4 marks]

b) State two other chemical stimuli that cause the heart rate to increase during exercise. [2 marks]

Q3 Atrial fibrillation (AF) is a condition that can result in a fast and irregular heartbeat because an abnormally high number of impulses are passed from the atria to the ventricles. Surgical treatment of AF can involve AVN ablation, which involves injuring the AVN so it no longer functions.

a) Suggest how this treatment helps to manage the condition. [2 marks]

b) After undergoing AVN ablation, patients also need to have a pacemaker implanted (an electronic device that sends out electrical impulses to control heart rate). Suggest why this is necessary. [2 marks]

My heart rate seems to be controlled by the boy next door...

It's also rising rapidly at the sight of so much to learn. You've got to properly learn it though — it's no good just having a rough idea. The SAN, baroreceptors, chemoreceptors — make sure you know what they are and what they do. Try drawing each row of the table above as a flow diagram, showing the route from stimulus to response.

Neurones

Ah, on to the good stuff. Notepad at the ready, motor neurones fired up, OK — lights, camera, action potentials...

Neurone **Cell Membranes** are **Polarised** at **Rest**

1) In a neurone's **resting state** (when it's not being stimulated), the **outside** of the membrane is **positively charged** compared to the **inside**. This is because there are **more positive ions outside** the cell than inside.
2) So the membrane is **polarised** — there's a **difference in charge** (called a **potential difference** or **voltage**) across it.
3) The voltage across the membrane when it's at rest is called the **resting potential** — it's about **–70 mV** (millivolts).
4) The resting potential is created and maintained by **sodium-potassium pumps** and **potassium ion channels** in a neurone's membrane:
   - The **sodium-potassium pumps** move **sodium ions out** of the neurone, but the membrane **isn't permeable** to **sodium ions**, so they **can't diffuse back in**. This creates a **sodium ion electrochemical gradient** (a **concentration gradient** of **ions**) because there are **more** positive sodium ions **outside** the cell than inside.
   - The sodium-potassium pumps also move **potassium ions in** to the neurone, but the membrane **is permeable** to **potassium ions** so they **diffuse back out** through **potassium ion channels**.
   - This makes the **outside** of the cell **positively charged** compared to the inside.

The sodium-potassium pump, potassium ion channel and sodium ion channel (see below) are all types of transport protein.

<u>Sodium-potassium pump</u> — These pumps use **active transport** to move **three sodium ions** (Na^+) **out** of the neurone for every **two potassium ions** (K^+) moved **in**. **ATP** is needed to do this.

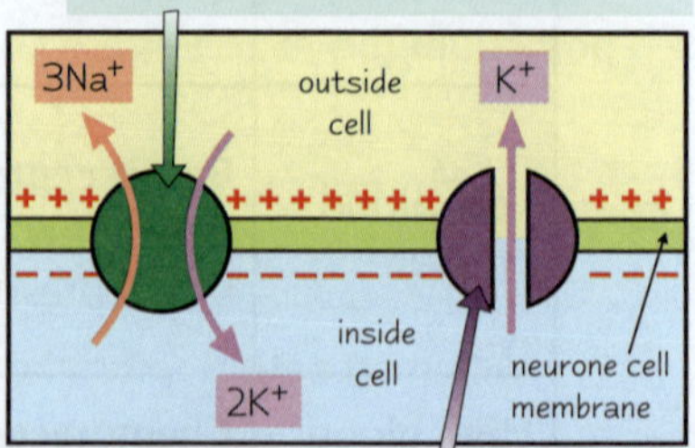

<u>Potassium ion channel</u> — These channels allow **facilitated diffusion** of **potassium ions** (K^+) **out** of the neurone, down their **concentration gradient**.

Neurone **Cell Membranes** Become **Depolarised** when They're **Stimulated**

A **stimulus** triggers other ion channels, called **sodium ion channels**, to **open**. If the stimulus is big enough, it'll trigger a **rapid change** in **potential difference**. The sequence of events is known as an **action potential**:

<u>Changes in potential difference during an action potential</u>

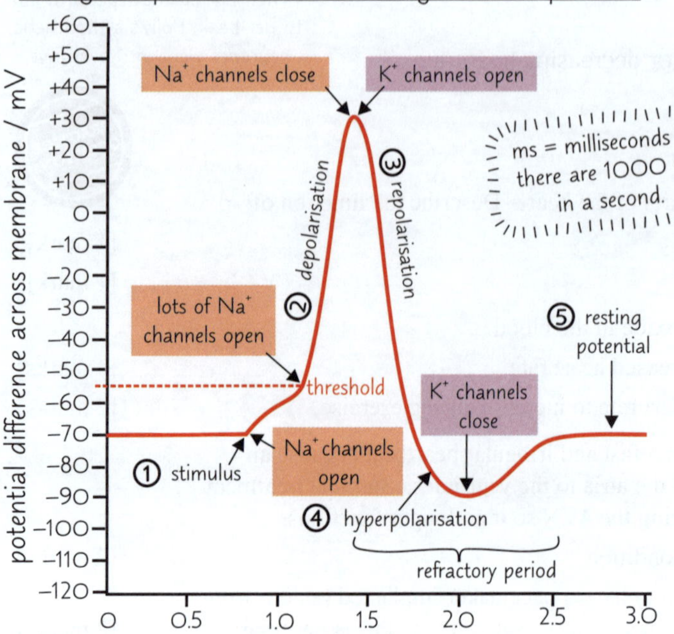

ms = milliseconds — there are 1000 ms in a second.

① **Stimulus** — this **excites** the neurone cell membrane, causing **sodium ion channels** to **open**. The membrane becomes **more permeable** to sodium, so **sodium ions diffuse into** the neurone down the sodium ion electrochemical gradient. This makes the **inside** of the neurone **less negative**.

② **Depolarisation** — if the potential difference reaches the **threshold** (around **–55 mV**), **more sodium ion channels open**. **More sodium ions diffuse rapidly into** the neurone.

③ **Repolarisation** — at a potential difference of around **+30 mV** the **sodium ion channels close** and **potassium ion channels open**. The membrane is **more permeable** to potassium so **potassium ions diffuse out** of the neurone down the potassium ion concentration gradient. This starts to get the membrane **back** to its **resting potential**.

The sodium channels have to close or the membrane will remain depolarised.

Neurones

④ **Hyperpolarisation** — **potassium ion channels** are **slow to close** so there's a slight '**overshoot**' where too many potassium ions diffuse out of the neurone. The potential difference becomes **more negative** than the **resting potential** (i.e. less than –70 mV).

⑤ **Resting potential** — the ion channels are **reset**. The **sodium-potassium pump** returns the membrane to its **resting potential** and maintains it until the membrane's excited by another stimulus.

After an **action potential**, the neurone cell membrane **can't** be **excited** again straight away. This is because the ion channels are **recovering** and they **can't** be made to **open** — **sodium ion channels** are **closed** during repolarisation and **potassium ion channels** are **closed** during hyperpolarisation. This period of recovery is called the **refractory period**.

The **Action Potential** Moves **Along** the **Neurone** as a **Wave** of **Depolarisation**

1) When an **action potential** happens, some of the **sodium ions** that enter the neurone **diffuse sideways**.
2) This causes **sodium ion channels** in the **next region** of the neurone to **open** and **sodium ions diffuse into** that part.
3) This causes a **wave of depolarisation** to travel along the neurone.
4) The **wave** moves **away** from the parts of the membrane in the **refractory period** because these parts **can't fire** an action potential.

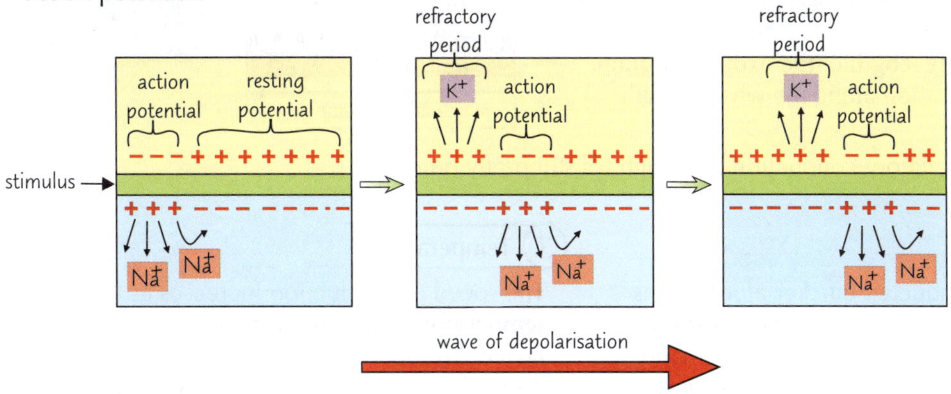

It's like a Mexican wave travelling through a crowd — sodium ions rushing inwards causes a wave of activity along the membrane.

The **Refractory Period** Produces **Discrete Impulses**

1) During the **refractory period**, **ion channels** are **recovering** and **can't** be **opened**.
2) So the refractory period acts as a **time delay** between one action potential and the next. This means that:
   - **action potentials don't overlap**, but pass along as **discrete** (separate) **impulses**.
   - there's a limit to the **frequency** at which the nerve impulses can be transmitted.
   - **action potentials** are **unidirectional** (they only travel in **one direction**).

Action Potentials have an **All-or-Nothing Nature**

1) Once the threshold is reached, an action potential will **always fire** with the **same change in voltage**, no matter how big the stimulus is.
2) If the **threshold isn't reached**, an action potential **won't fire**. This is the **all-or-nothing** nature of action potentials.
3) A **bigger stimulus** won't cause a bigger action potential, but it will cause them to fire **more frequently**.

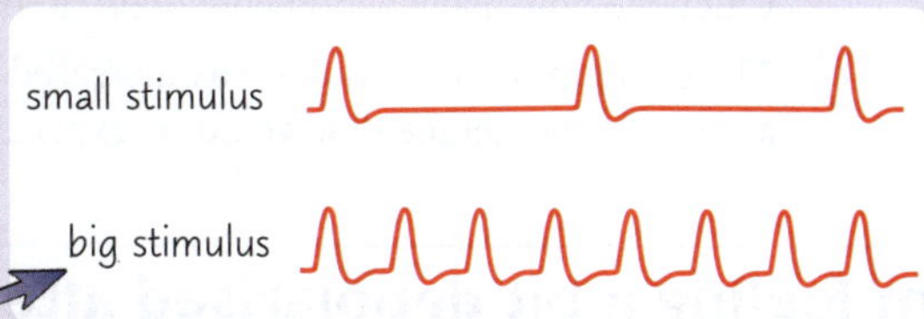

Neurones

Three Factors Affect the Speed of Conduction of Action Potentials

1 Myelination

1) Some neurones are **myelinated** — they have a **myelin sheath**.
2) The myelin sheath is an **electrical insulator**.
3) In the peripheral nervous system, the sheath is made of a type of cell called a **Schwann cell**.
4) Between the Schwann cells are tiny patches of **bare membrane** called the **nodes of Ranvier**. **Sodium ion channels** are **concentrated** at the nodes.
5) In a **myelinated** neurone, **depolarisation** only happens at the **nodes of Ranvier** (where sodium ions can get through the membrane).
6) The neurone's **cytoplasm conducts** enough electrical charge to **depolarise** the **next node**, so the impulse '**jumps**' from node to node.
7) This is called **saltatory conduction** and it's **really fast**.
8) In a **non-myelinated** neurone, the impulse travels as a **wave** along the **whole length** of the **axon membrane** (so you get **depolarisation** along the whole length of the membrane).
9) This is **slower** than saltatory conduction (although it's still pretty quick).

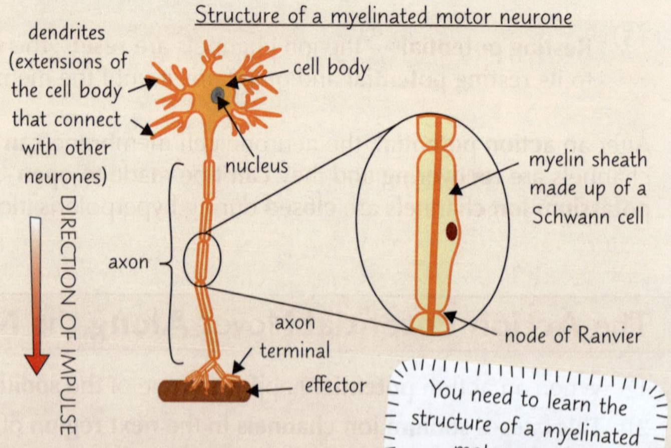

You need to learn the structure of a myelinated motor neurone.

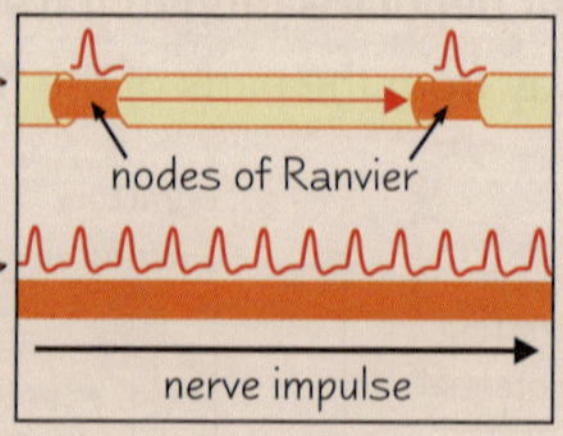

2 Axon diameter

Action potentials are conducted **quicker** along axons with **bigger diameters** because there's **less resistance** to the **flow of ions** than in the cytoplasm of a smaller axon. With less resistance, **depolarisation reaches** other parts of the neurone cell membrane **quicker**.

3 Temperature

The speed of conduction increases as the **temperature increases** too, because **ions diffuse faster**. The speed only increases up to around **40 °C** though — after that the **proteins** begin to **denature** and the speed decreases.

Warm-Up Questions

PRACTICE QUESTIONS

Q1 Give one function of the refractory period.

Q2 What is meant by the 'all-or-nothing' nature of action potentials?

Q3 What is the function of Schwann cells on a neurone?

Q4 Give three factors that affect the speed of conduction of action potentials.

Exam Question

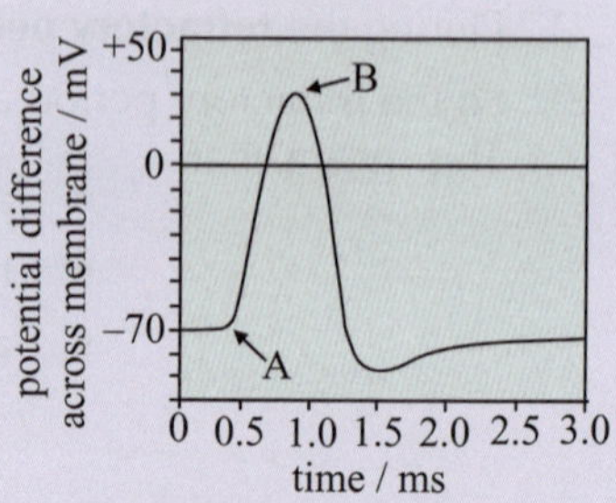

Q1 The graph shows an action potential across an axon membrane following the application of a stimulus.

a) Explain what causes the change in potential difference between point A and point B. [2 marks]

b) The same stimulus was applied consistently for over one hour. The next action potential fired at 4.5 ms. Calculate how many action potentials fired in one hour. Give your answer in standard form. [2 marks]

c) The strength of the stimulus was increased by 50%. Give the maximum potential difference across the membrane that would be experienced with this stronger stimulus. [1 mark]

I'm feeling a bit depolarised after all that...

All this stuff about neurones can be a bit tricky to get your head around. Take your time and try scribbling it all down a few times till it starts to make some kind of sense. Neurones work because there's an electrical charge across their membrane, which is set up by ion pumps and ion channels. It's a change in this charge that transmits an action potential.

Synaptic Transmission

When an action potential arrives at the end of a neurone, the information has to be passed on to the next cell — this could be another neurone, a muscle cell or a gland cell.

A **Synapse** is a **Junction** Between a **Neurone** and the **Next Cell**

Typical structure of a synapse

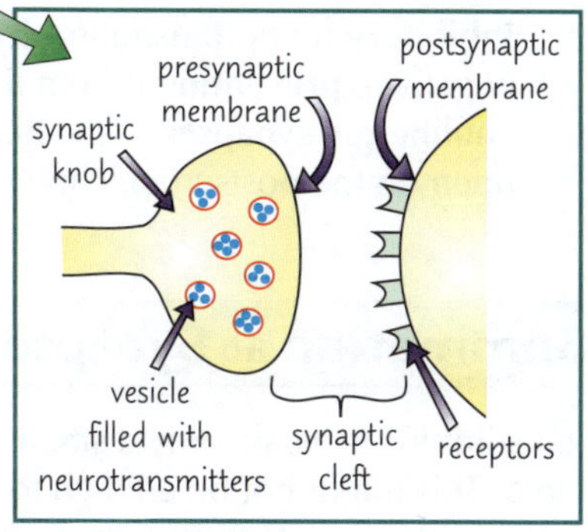

1) A **synapse** is the junction between a **neurone** and another **neurone**, or between a **neurone** and an **effector cell**, e.g. a muscle or gland cell.
2) The **tiny gap** between the cells at a synapse is called the **synaptic cleft**.
3) The **presynaptic neurone** (the one before the synapse) has a **swelling** called a **synaptic knob**. This contains **synaptic vesicles** filled with **chemicals** called **neurotransmitters**.
4) When an **action potential** reaches the end of a neurone it causes **neurotransmitters** to be **released** into the synaptic cleft. They **diffuse across** to the **postsynaptic membrane** (the one after the synapse) and **bind** to **specific receptors**.
5) When neurotransmitters bind to receptors they might **trigger** an **action potential** (in a neurone), cause **muscle contraction** (in a muscle cell), or cause a **hormone** to be **secreted** (from a gland cell).
6) Because the receptors are **only** on the postsynaptic membranes, synapses make sure impulses are **unidirectional** — the impulse can only travel in **one direction**.
7) Neurotransmitters are **removed** from the **cleft** so the **response** doesn't keep happening, e.g. they're taken back into the **presynaptic neurone** or they're **broken down** by **enzymes** (and the products are taken into the neurone).
8) There are many **different** neurotransmitters, e.g. **acetylcholine** (**ACh**) and **noradrenaline**. Synapses that use acetylcholine are called **cholinergic synapses**. Their structure is exactly the **same** as in the diagram above.

ACh Transmits the Nerve Impulse **Across** a **Cholinergic Synapse**

This is how a **nerve impulse** is transmitted across a **cholinergic synapse**:

1) An action potential (see pages 36-37) arrives at the **synaptic knob** of the **presynaptic neurone**.
2) The action potential stimulates **voltage-gated calcium ion channels** in the **presynaptic neurone** to **open**.
3) **Calcium ions diffuse into** the synaptic knob. (They're pumped out afterwards by active transport.)

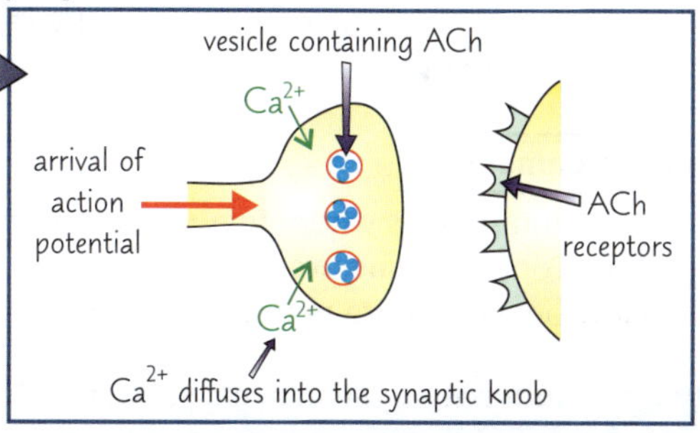

Voltage-gated ion channels open at a certain voltage.

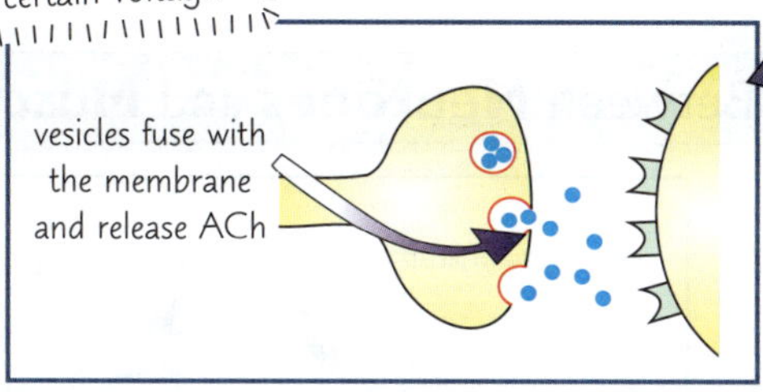

4) The influx of **calcium ions** into the synaptic knob causes the **synaptic vesicles** to **move** to the **presynaptic membrane**. They then **fuse** with the presynaptic membrane.
5) The **vesicles release** the neurotransmitter **acetylcholine** (**ACh**) into the **synaptic cleft** — this is called **exocytosis**.

6) ACh **diffuses** across the **synaptic cleft** and **binds** to specific **cholinergic receptors** on the **postsynaptic membrane**.
7) This causes **sodium ion channels** in the **postsynaptic neurone** to **open**.
8) The **influx** of **sodium ions** into the postsynaptic membrane causes **depolarisation**. An **action potential** on the **postsynaptic membrane** is generated if the **threshold** is reached.
9) ACh is **removed** from the **synaptic cleft** so the **response** doesn't keep happening. It's **broken down** by an **enzyme** called **acetylcholinesterase** (**AChE**) and the products are **re-absorbed** by the **presynaptic neurone** and used to make more ACh.

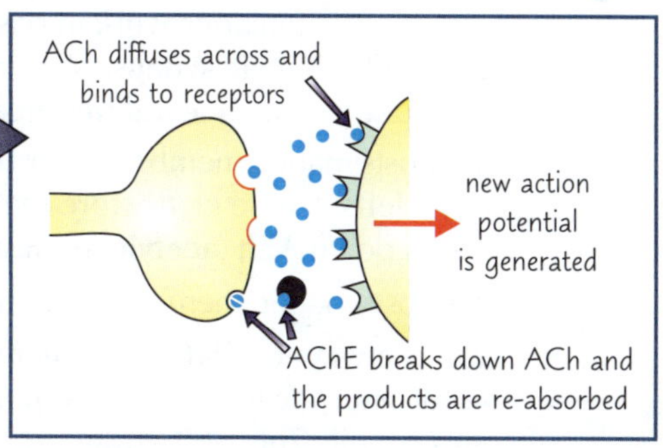

Synaptic Transmission

Neurotransmitters Can be Excitatory, Inhibitory or Both

1) **Excitatory** neurotransmitters **depolarise** the postsynaptic membrane, making it fire an **action potential** if the **threshold** is reached. E.g. **acetylcholine** is an excitatory neurotransmitter at **cholinergic synapses** in the **CNS** — it binds to cholinergic receptors to cause an **action potential** in the postsynaptic membrane — and at **neuromuscular junctions** (see below).
2) **Inhibitory** neurotransmitters **hyperpolarise** the postsynaptic membrane (make the potential difference more negative), **preventing** it from firing an action potential. E.g. **acetylcholine** is an **inhibitory** neurotransmitter at **cholinergic synapses** in the **heart**. When it binds to receptors here, it can cause **potassium ion channels** to **open** on the postsynaptic membrane, **hyperpolarising** it.

Summation at Synapses Finely Tunes the Nervous Response

If a stimulus is **weak**, only a **small amount** of **neurotransmitter** will be released from a neurone into the synaptic cleft. This might not be enough to **excite** the postsynaptic membrane to the **threshold** level and stimulate an action potential. **Summation** is where the effect of neurotransmitter released from many neurones (or one neurone that's stimulated a lot in a short period of time) is **added together**. There are two types of summation:

Spatial summation

1) Sometimes **many** neurones **connect** to **one** neurone.
2) The small amount of **neurotransmitter** released from **each** of these neurones can be enough **altogether** to **reach** the **threshold** in the postsynaptic neurone and **trigger** an **action potential**.
3) If some neurones release an **inhibitory neurotransmitter** then the total effect of all the neurotransmitters might be **no action potential**.

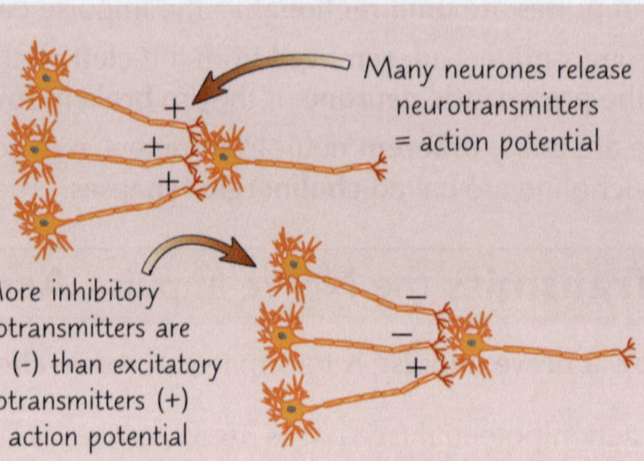

Temporal summation

Temporal summation is where **two or more** nerve impulses arrive in **quick succession** from the **same presynaptic neurone**. This makes an action potential **more likely** because **more neurotransmitter** is released into the **synaptic cleft**.

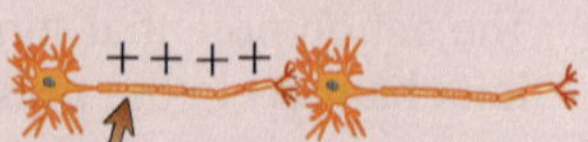

Both types of **summation** mean synapses **accurately process information**, **finely tuning** the response.

Neuromuscular Junctions are Synapses Between Neurones and Muscles

1) A **neuromuscular junction** is a **synapse** between a **motor neurone** and a **muscle cell**.
2) Neuromuscular junctions use the neurotransmitter **acetylcholine** (**ACh**), which binds to cholinergic receptors called **nicotinic cholinergic receptors**.
3) Neuromuscular junctions **work** in basically the **same way** as the **cholinergic synapse** shown on the previous page — but there are a few **differences**:
   - The postsynaptic membrane has lots of **folds** that form **clefts**. These clefts **store** the **enzyme** that breaks down **ACh** (**acetylcholinesterase** — **AChE**).
   - The postsynaptic membrane has **more receptors** than other synapses.
   - ACh is **always excitatory** at a **neuromuscular junction**. So when a **motor neurone** fires an **action potential**, it normally triggers a **response** in a muscle cell. This **isn't** always the case for a synapse between two neurones.

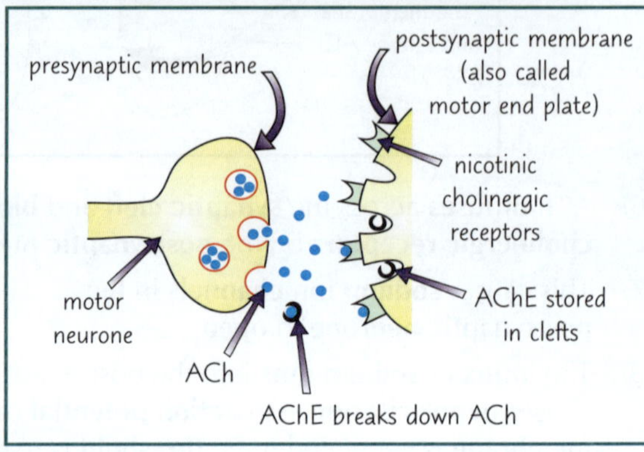

You need to be able to compare transmission across a cholinergic synapse and a neuromuscular junction.

Synaptic Transmission

Drugs Affect the Action of Neurotransmitters at Synapses in Various Ways

Some **drugs affect synaptic transmission**. You might have to **predict** the **effects** that a drug would have at a synapse in your exam. Here are some **examples** of how drugs can affect synaptic transmission:

You don't need to learn the names of the drugs.

1. Some drugs are the **same shape** as neurotransmitters so they **mimic** their action at receptors (these drugs are called **agonists**). This means **more receptors** are **activated**. E.g. **nicotine** mimics **acetylcholine** so binds to nicotinic cholinergic receptors in the brain.

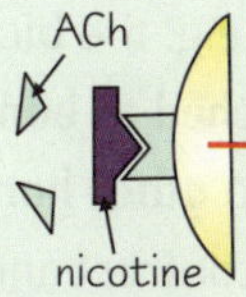

2. Some drugs **block receptors** so they **can't be activated** by neurotransmitters (these drugs are called **antagonists**). This means **fewer receptors** (if any) can be **activated**. E.g. **curare** blocks the effects of acetylcholine by blocking nicotinic cholinergic receptors at neuromuscular junctions, so muscle cells can't be stimulated. This results in the muscle being **paralysed**.

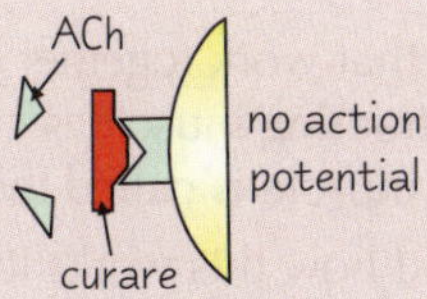

3. Some drugs **inhibit** the **enzyme** that breaks down neurotransmitters (they stop it from working). This means there are **more neurotransmitters** in the synaptic cleft to **bind** to **receptors** and they're there for **longer**. E.g. **nerve gases** stop acetylcholine from being broken down in the synaptic cleft. This can lead to **loss** of **muscle control**.

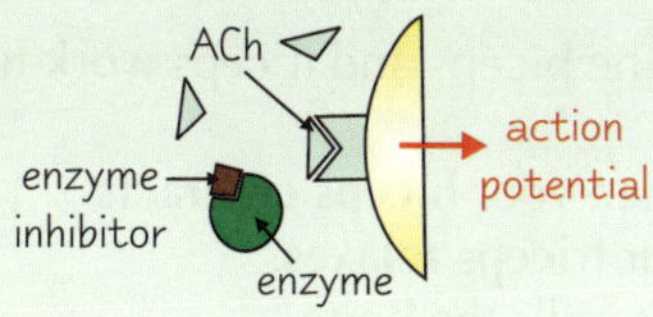

4. Some drugs **stimulate** the release of **neurotransmitter** from the presynaptic neurone so **more receptors** are activated, e.g. **amphetamines**.

5. Some drugs **inhibit** the release of neurotransmitters from the presynaptic neurone so **fewer receptors** are activated, e.g. **alcohol**.

Warm-Up Questions

PRACTICE QUESTIONS

Q1 How do synapses ensure that nerve impulses are unidirectional?

Q2 Give one way that neurotransmitters are removed from the synaptic cleft.

Q3 Which neurotransmitter do you find at cholinergic synapses?

Q4 Why are calcium ions important in synaptic transmission?

Q5 What do inhibitory neurotransmitters do at synapses?

Q6 What kind of receptors are found at neuromuscular junctions?

Exam Questions

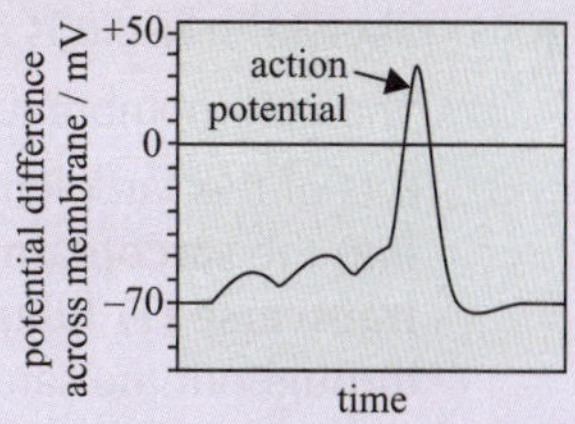

Q1 The graph on the right shows the potential difference across a postsynaptic membrane against time.

a) Suggest why a potential difference of –45 mV is significant for this postsynaptic membrane. *[1 mark]*

b) The action potential shown on the graph was fired as a result of temporal summation. Use the graph and your own knowledge to explain how this action potential was created. *[4 marks]*

Q2 Myasthenia gravis is a disease in which the body's immune system gradually destroys receptors at neuromuscular junctions. This leads to weaker muscular responses than normal. Explain why. *[3 marks]*

Q3 Galantamine is a drug that inhibits the enzyme acetylcholinesterase (AChE). Predict the effect of galantamine at a neuromuscular junction and explain your answer. *[3 marks]*

Neurotransmitter revision inhibits any excitement...

Some more pretty tough pages here — lovely. And lots more diagrams to have a go at drawing and re-drawing. Don't worry if you're not the world's best artist, just make sure you add labels to your drawings to explain what's happening.

Muscle Contraction

I reckon muscle cells are the spoilt brats of the Biology world. They're so special that everything muscly has to have its own special name — there's none of this "cell membrane" malarkey, oh no, it's "sarcolemma" if you please...

Muscles Act in **Antagonistic Pairs**

1) **Skeletal muscle** (also called striated, striped or voluntary muscle) is the type of muscle you use to **move**, e.g. the biceps and triceps move the lower arm.
2) Skeletal muscles are **attached** to **bones** by **tendons**.
3) **Ligaments attach bones** to **other bones**, to hold them together.
4) Pairs of skeletal muscles **contract** and **relax** to **move bones** at a **joint**. The bones of the skeleton are **incompressible** (rigid) so they act as **levers**, giving the muscles something to **pull** against.
5) Muscles that work together to move a bone are called **antagonistic pairs**. The **contracting** muscle is called the **agonist** and the **relaxing** muscle is called the **antagonist**.

Muscles work in pairs because they can only pull when they contract — they can't push.

To understand how this works it's best to look at an example:

- The bones of your **lower arm** are attached to a **biceps** muscle and a **triceps** muscle by **tendons**.
- The biceps and triceps **work together** to move your arm — as one **contracts**, the other **relaxes**.

When your **biceps contracts** your **triceps relaxes**. This pulls the bone so your **arm bends** (**flexes**) at the elbow. Here, the **biceps** is the **agonist** and the **triceps** is the **antagonist**.

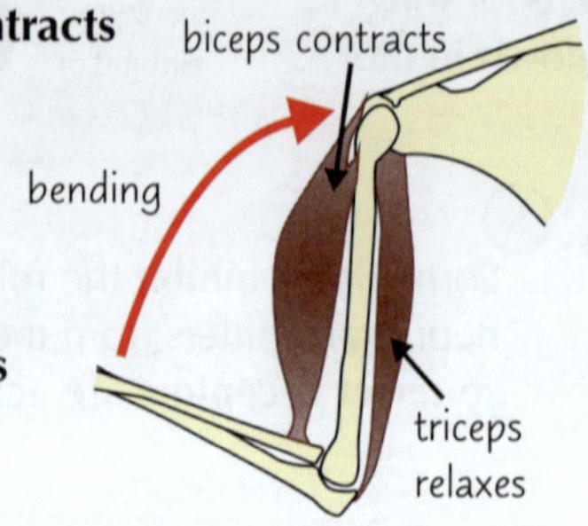

When your **triceps contracts** your **biceps relaxes**. This pulls the bone so your **arm straightens** (**extends**) at the **elbow**. Here, the **triceps** is the **agonist** and the **biceps** is the **antagonist**.

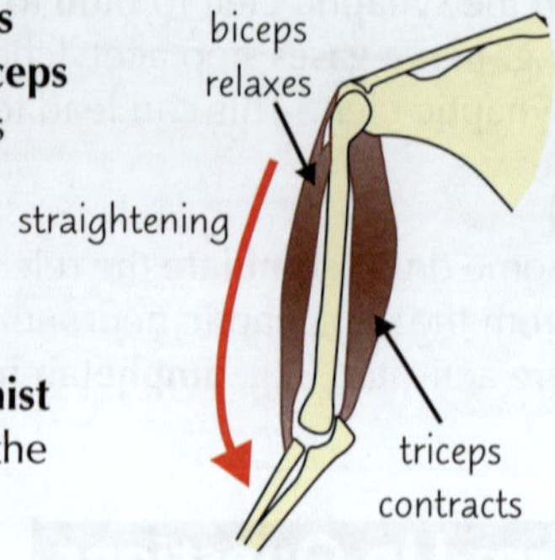

Skeletal Muscle is Made Up of **Long Muscle Fibres**

Muscles act as **effectors** and are **stimulated** to **contract** by neurones.

1) Skeletal muscle is made up of **large bundles** of **long cells**, called **muscle fibres**.
2) The cell membrane of muscle fibre cells is called the **sarcolemma**.
3) Bits of the sarcolemma **fold inwards** across the muscle fibre and stick into the **sarcoplasm** (a muscle cell's cytoplasm). These folds are called **transverse (T) tubules** and they help to **spread electrical impulses** throughout the sarcoplasm so they **reach** all parts of the **muscle fibre**.
4) A network of **internal membranes** called the **sarcoplasmic reticulum** runs through the sarcoplasm. The sarcoplasmic reticulum **stores** and **releases calcium ions** that are needed for muscle contraction (see p. 44).
5) Muscle fibres have lots of **mitochondria** to **provide** the **ATP** that's needed for **muscle contraction**.
6) Muscle fibres are **multinucleate** (contain many nuclei).
7) Muscle fibres have lots of **long**, **cylindrical organelles** called **myofibrils**. They're made up of proteins and are **highly specialised** for **contraction**.

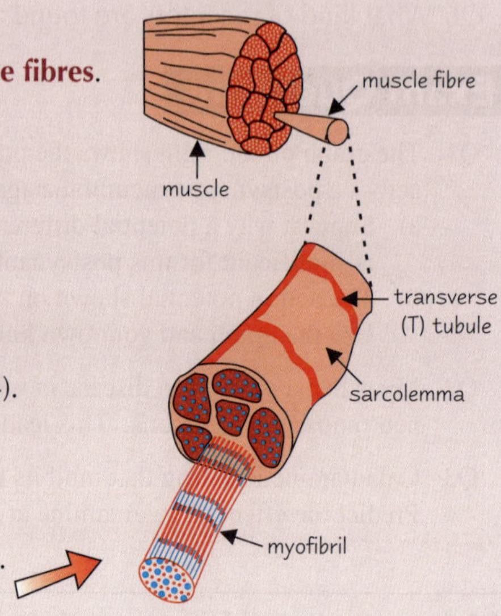

Muscle Contraction

Myofibrils Contain Thick Myosin Filaments and Thin Actin Filaments

1) Myofibrils contain bundles of **thick** and **thin myofilaments** that **move past each other** to make muscles **contract**.
   - **Thick myofilaments** are made of the protein **myosin**.
   - **Thin myofilaments** are made of the protein **actin**.
2) If you look at a **myofibril** under an electron **microscope**, you'll see a pattern of alternating **dark** and **light bands**:
   - **Dark** bands contain the **thick myosin filaments** and some overlapping thin actin filaments — these are called **A-bands**.
   - **Light** bands contain **thin actin filaments** only — these are called **I-bands**.
3) A myofibril is made up of many short units called **sarcomeres**.
4) The **ends** of each **sarcomere** are marked with a **Z-line**.
5) In the **middle** of each sarcomere is an **M-line**. The M-line is the **middle** of the **myosin** filaments.
6) **Around** the M-line is the **H-zone**. The H-zone **only** contains **myosin** filaments.

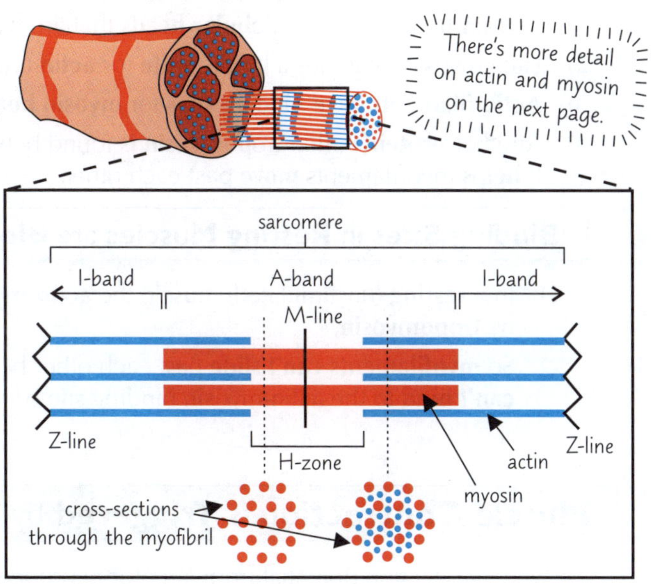

Muscle Contraction is Explained by the Sliding Filament Theory

1) **Myosin** and **actin** filaments **slide** over one another to make the **sarcomeres contract** — the myofilaments themselves **don't** contract.
2) The **simultaneous contraction** of lots of **sarcomeres** means the **myofibrils** and **muscle fibres contract**.
3) Sarcomeres return to their **original length** as the muscle **relaxes**.

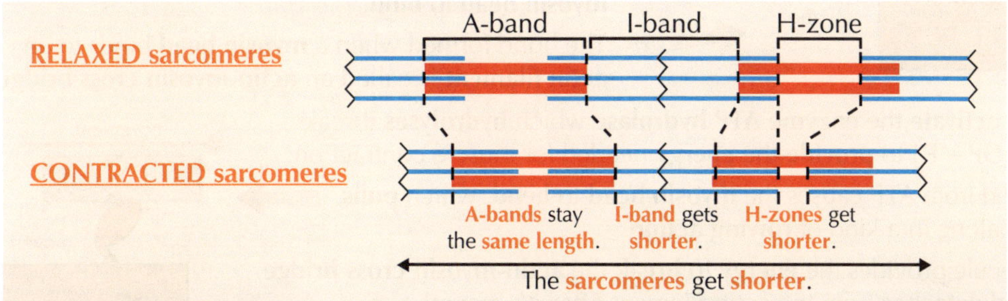

If only the sliding filament theory was as much fun...

Warm-Up Questions

Q1 Describe one example of how muscles act in antagonistic pairs.

Q2 What are transverse (T) tubules?

Q3 Name the two proteins that make up myofibrils.

Exam Question

Q1 A muscle myofibril was examined under an electron microscope and a sketch was drawn (Figure 1).

Figure 1

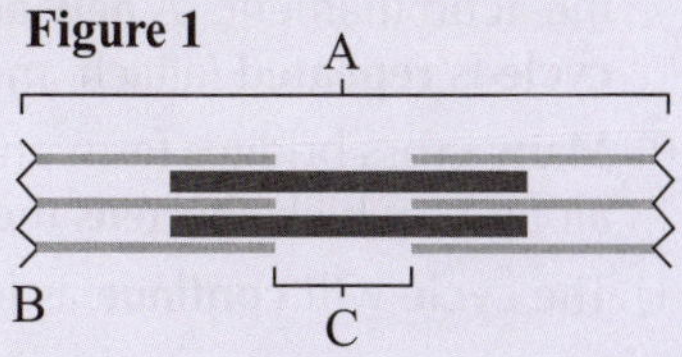

Figure 2

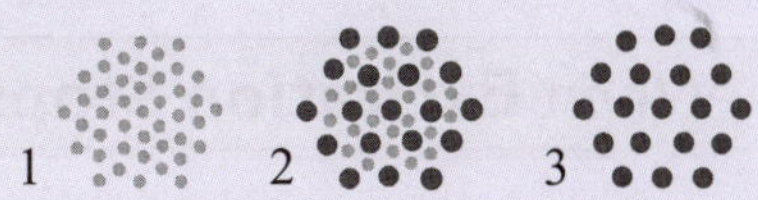

a) What are the correct names for labels A, B and C? [3 marks]

b) Describe how the lengths of the different bands in a myofibril change during muscle contraction. [2 marks]

c) The myofibril was then cut through the M-line (Figure 2). State which of the cross-section drawings you would expect to see and explain why. [3 marks]

Sarcomere — a French mother with a dry sense of humour...

Blimey, there are an awful lot of similar-sounding names to learn on these pages. And then you've got your A-band, I-band, what-band, who-band to memorise too. But once you've learnt them, these are things you'll never forget.

Muscle Contraction

Myofilaments sliding over one another takes a lot of energy — probably why exercise is such hard work...

Myosin Filaments Have Globular Heads and Binding Sites

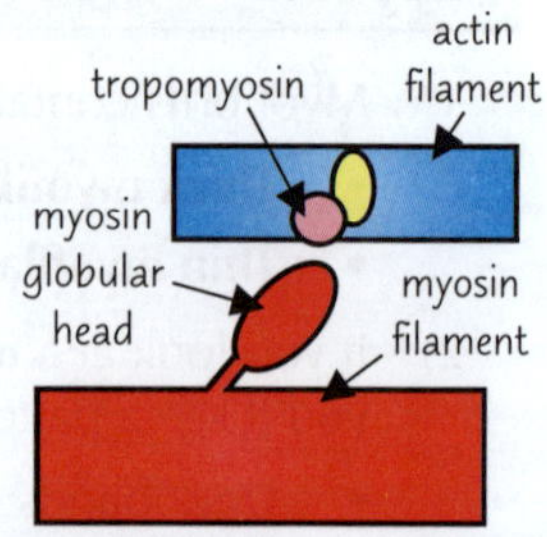

1) **Myosin filaments** have **globular heads** that are **hinged**, so they can move **back** and **forth**.
2) Each myosin head has a **binding site** for **actin** and a **binding site** for **ATP**.
3) **Actin filaments** have **binding sites** for **myosin heads**, called **actin-myosin** binding sites.
4) Another **protein** called **tropomyosin** is found between actin filaments. It **helps** myofilaments **move** past each other.

Binding Sites in Resting Muscles are Blocked by Tropomyosin

1) In a **resting** (unstimulated) muscle the **actin-myosin binding site** is **blocked** by **tropomyosin**.
2) So **myofilaments can't slide** past each other because the **myosin heads can't bind** to the actin-myosin binding site on the actin filaments.

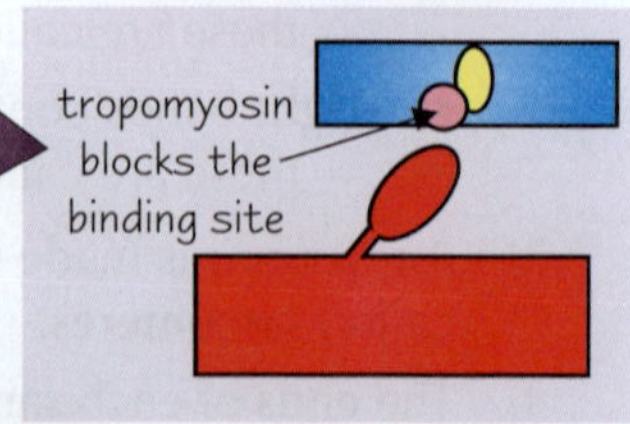

Muscle Contraction is Triggered by an Influx of Calcium Ions

1) When an action potential from a motor neurone **stimulates** a muscle cell, it **depolarises** the **sarcolemma**. Depolarisation **spreads** down the **T-tubules** to the **sarcoplasmic reticulum** (see p. 42).
2) This causes the **sarcoplasmic reticulum** to **release** stored **calcium ions** (Ca^{2+}) into the **sarcoplasm**.
3) Calcium ions **bind** to a protein attached to tropomyosin, causing the protein to **change shape**. This **pulls** the attached **tropomyosin out** of the **actin-myosin binding site** on the actin filament.
4) This **exposes** the **binding site**, which allows the **myosin head** to **bind**.
5) The bond formed when a **myosin head** binds to an **actin filament** is called an **actin-myosin cross bridge**.

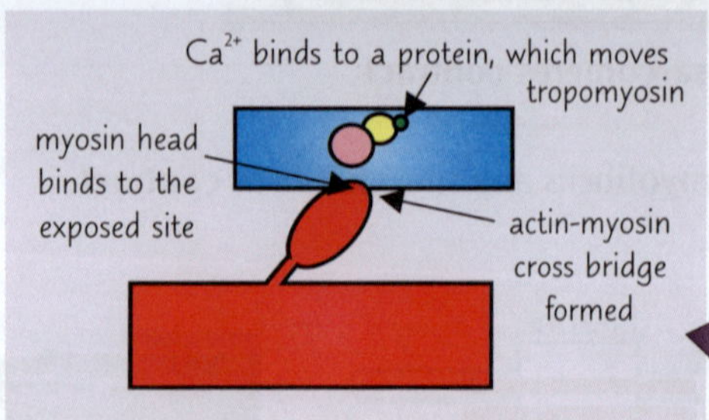

6) **Calcium** ions also **activate** the enzyme **ATP hydrolase** which **hydrolyses** (breaks down) **ATP** (into ADP + P_i) to **provide** the **energy** needed for muscle contraction.
7) The **energy** released from ATP causes the **myosin head** to **bend**, which **pulls** the **actin filament** along in a kind of **rowing action**.
8) Another **ATP** molecule provides the **energy** to **break** the **actin-myosin cross bridge**, so the **myosin head detaches** from the actin filament **after** it's moved.
9) The **myosin head** then **reattaches** to a **different binding site** further along the actin filament. A **new actin-myosin cross bridge** is formed and the **cycle** is **repeated** (attach, move, detach, reattach to new binding site...).
10) **Many** cross bridges **form** and **break** very **rapidly**, pulling the actin filament along — which **shortens** the **sarcomere**, causing the **muscle** to **contract**.
11) The cycle will **continue** as long as **calcium ions** are **present**.

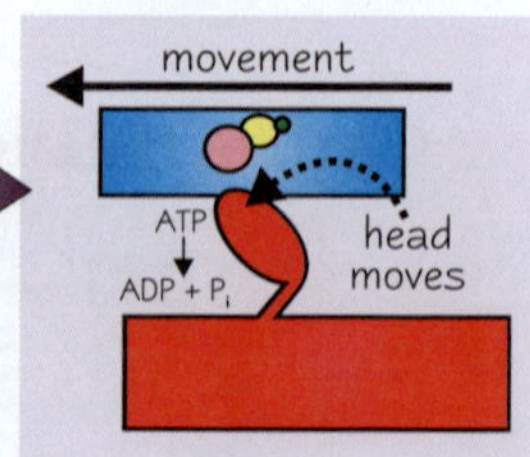

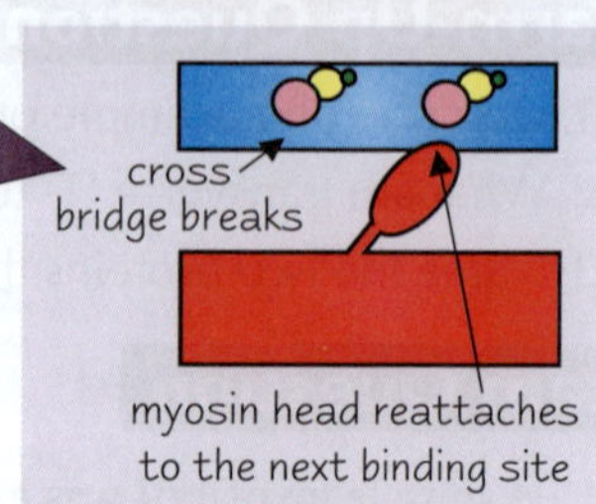

When Excitation Stops, Calcium Ions Leave

1) When the muscle **stops** being **stimulated**, **calcium ions leave** their **binding sites** and are moved by **active transport** back into the **sarcoplasmic reticulum** (this needs **ATP** too).
2) This causes the **tropomyosin** molecules to **move back**, so they **block** the **actin-myosin binding sites** again.
3) Muscles **aren't contracted** because **no myosin heads** are **attached** to **actin** filaments (so there are no actin-myosin cross bridges).
4) The **actin** filaments **slide back** to their **relaxed** position, which **lengthens** the **sarcomere**.

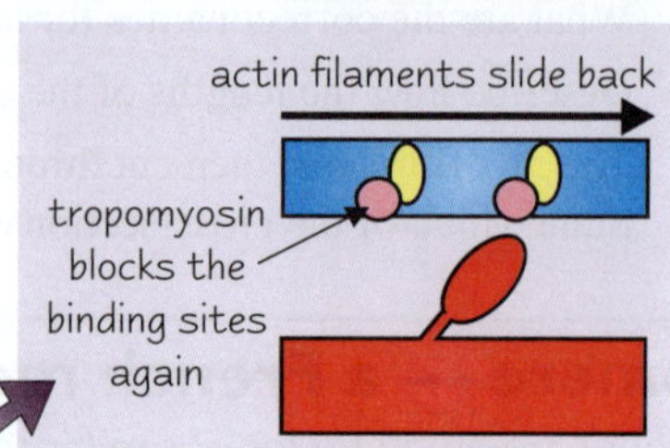

Muscle Contraction

ATP and Phosphocreatine Provide the Energy for Muscle Contraction

So much **energy** is **needed** when muscles contract that **ATP** gets **used up very quickly**.
ATP has to be **continually generated** so exercise can continue — this happens in **three main ways**:

1) Aerobic respiration
   - **Most ATP** is generated via **oxidative phosphorylation** in the cell's **mitochondria**.
   - **Aerobic** respiration only works when there's **oxygen** so it's good for **long periods** of **low-intensity exercise**.

2) Anaerobic respiration
   - ATP is made **rapidly** by **glycolysis**.
   - The **end product** of glycolysis is **pyruvate**, which is converted to **lactate** by **lactate fermentation**.
   - Lactate can **quickly build up** in the muscles and cause **muscle fatigue**.
   - Anaerobic respiration is good for **short periods** of **hard exercise**, e.g. a **400 m sprint**.

See pages 12-15 for more on aerobic and anaerobic respiration.

3) ATP-Phosphocreatine (PCr) System
   - **ATP** is made by **phosphorylating ADP** — adding a phosphate group taken from **PCr**.

     ADP + **PCr** → **ATP** + **Cr** (creatine)

   - **PCr** is **stored** inside cells and the ATP-PCr system **generates ATP** very **quickly**.
   - **PCr runs out** after a few seconds so it's used during **short bursts** of **vigorous exercise**, e.g. a **tennis serve**.
   - The ATP-PCr system is **anaerobic** (it doesn't need oxygen) and it's **alactic** (it doesn't form any lactate).

Many activities use a combination of these systems.

Some of the creatine (Cr) gets broken down into **creatinine**, which is removed from the body via the **kidneys**. Creatinine levels can be higher in people who **exercise regularly** and those with a **high muscle mass**. High creatinine levels may also indicate **kidney damage**.

Skeletal Muscles are Made of Slow Twitch and Fast Twitch Muscle Fibres

Skeletal muscles are made up of **two types** of **muscle fibres** — **slow twitch** and **fast twitch**.
Different muscles have **different proportions** of slow and fast twitch fibres. The two types have **different properties**:

SLOW TWITCH MUSCLE FIBRES	FAST TWITCH MUSCLE FIBRES
Muscle fibres that contract slowly.	Muscle fibres that contract very quickly.
Muscles you use for posture, e.g. those in the back, have a high proportion of them.	Muscles you use for fast movement, e.g. those in the eyes and legs, have a high proportion of them.
Good for endurance activities, e.g. maintaining posture, long-distance running.	Good for short bursts of speed and power, e.g. eye movement, sprinting.
Can work for a long time without getting tired.	Get tired very quickly.
Energy's released slowly through aerobic respiration. Lots of mitochondria and blood vessels supply the muscles with oxygen.	Energy's released quickly through anaerobic respiration using glycogen (stored glucose). There are few mitochondria or blood vessels.
Reddish in colour because they're rich in myoglobin — a red-coloured protein that stores oxygen.	Whitish in colour because they don't have much myoglobin (so can't store much oxygen).

Warm-Up Questions

Q1 Describe one way that ATP can be generated in contracting muscles.

Q2 State three differences between slow and fast twitch skeletal muscle fibres.

Exam Questions

Q1 Rigor mortis is the stiffening of muscles in the body after death. It happens when ATP reserves are exhausted. Explain why a lack of ATP leads to muscles being unable to relax. [3 marks]

Q2 Bepridil is a drug that blocks calcium ion channels.
Describe and explain the effect this drug will have on muscle contraction. [3 marks]

What does muscle contraction cost? 80p...

Sorry, that's my favourite sciencey joke so I had to fit it in somewhere — a small distraction before you revisit this page. It's tough stuff but you know the best way to learn it. That's right, grab yourself a nice felt-tip pen and a pad of paper...

Homeostasis Basics

Ah, there's nothing like learning a nice long word to start you off on a new section — welcome to homeostasis.

Homeostasis is the **Maintenance** of a **Stable Internal Environment**

1) **Changes** in your **external environment** can affect your **internal environment** — the blood and tissue fluid that surrounds your cells.
2) **Homeostasis** involves **control systems** that keep your **internal environment** roughly **constant** (within **certain limits**).
3) **Keeping** your internal environment **stable** is vital for cells to **function normally** and to **stop** them being **damaged**.
4) It's particularly important to **maintain** the right **core body temperature** and **blood pH**. This is because temperature and pH affect **enzyme activity**, and enzymes **control** the **rate** of **metabolic reactions**:

Temperature

- If body temperature is **too high** (e.g. 40 °C) **enzymes** may become **denatured**. The enzyme's molecules **vibrate too much**, which **breaks** the **hydrogen bonds** that hold them in their **3D shape**. The **shape** of the enzyme's **active site** is **changed** and it **no longer works** as a **catalyst**. This means **metabolic reactions** are **less efficient**.
- If body temperature is **too low enzyme activity** is **reduced**, **slowing** the rate of **metabolic reactions**.
- The **highest rate** of **enzyme activity** happens at their **optimum temperature** (about **37 °C** in humans).

pH

- If blood pH is **too high** or **too low** (highly alkaline or acidic) **enzymes** become **denatured**. The **hydrogen bonds** that hold them in their 3D shape are broken, so the **shape** of the enzyme's **active site** is **changed** and it **no longer works** as a **catalyst**. This means **metabolic reactions** are **less efficient**.
- The **highest rate** of **enzyme activity** happens at their **optimum pH** — usually **around pH 7** (neutral), but some enzymes work best at other pHs, e.g. enzymes found in the stomach work best at a low pH.

5) It's important to **maintain** the right **concentration** of **glucose** in the **blood** because cells need glucose for **energy**. Blood glucose concentration also affects the **water potential** of blood — this is the potential (likelihood) of water molecules to **diffuse** out of or into a solution.

Glucose

- If blood glucose concentration is **too high** the **water potential** of blood is **reduced** to a point where **water** molecules **diffuse out** of cells into the blood by osmosis. This can cause the cells to **shrivel up** and **die**.
- If blood glucose concentration is **too low**, cells are **unable** to carry out **normal activities** because there **isn't enough glucose** for **respiration** to provide **energy**.

Homeostatic Systems Detect a **Change** and **Respond** by **Negative Feedback**

1) Homeostatic systems involve **receptors**, a **communication system** and **effectors** (like the nervous system — see page 28).
2) Receptors detect when a level is **too high** or **too low**, and the information's communicated via the **nervous** system or the **hormonal** system to **effectors**.
3) The effectors respond to **counteract** the change — bringing the level **back** to **normal**.
4) The mechanism that **restores** the level to **normal** is called a **negative feedback** mechanism.
5) Negative feedback **keeps** things around the **normal** level, e.g. body temperature is usually kept **within 0.5 °C** above or below **37 °C**.
6) Negative feedback only works within **certain limits** though — if the change is **too big** then the **effectors** may **not** be able to **counteract** it, e.g. a huge drop in body temperature caused by prolonged exposure to cold weather may be too large to counteract.

Control of **body temperature** by **negative feedback**:

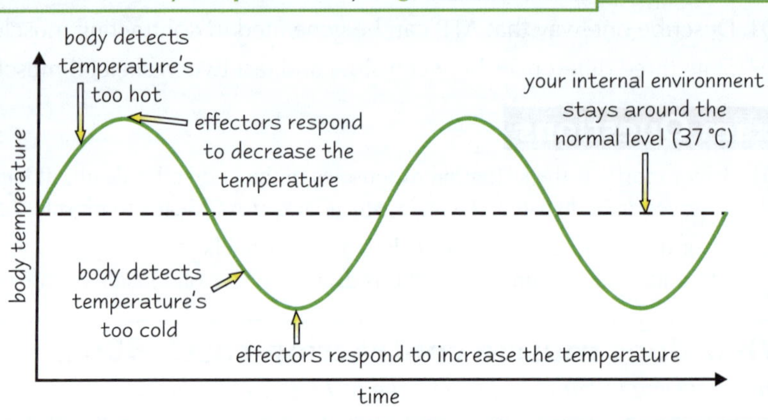

Homeostasis Basics

Multiple Negative Feedback Mechanisms Give More Control

1) Homeostasis involves **multiple negative feedback mechanisms** for each thing being controlled. This is because having more than one mechanism gives **more control** over changes in your internal environment than just having one negative feedback mechanism.
2) Having multiple negative feedback mechanisms means you can **actively increase** or **decrease a level** so it returns to **normal**, e.g. you have feedback mechanisms to reduce your body temperature and you also have mechanisms to increase it.
3) If you only had **one negative feedback mechanism**, all you could do would be **turn it on** or **turn it off**. You'd only be able to actively change a level in **one direction** so it returns to normal, e.g. it's a bit like trying to slow down a car with only an accelerator — all you can do is take your foot off the accelerator (you'd have more control with a brake too).
4) Only **one** negative feedback mechanism means a **slower response** and **less control**.

There was plenty of negative feedback when Carl wore his new vest-pants combo out for dinner.

Positive Feedback Mechanisms Amplify a Change from the Normal Level

1) Some changes trigger a **positive feedback** mechanism, which **amplifies** the change.
2) The effectors respond to **further increase** the level **away** from the **normal** level.
3) Positive feedback is useful to **rapidly activate** something, e.g. a **blood clot** after an injury.

- **Platelets** become **activated** and release a **chemical** — this triggers **more platelets** to be activated, and so on.
- Platelets **very quickly** form a **blood clot** at the injury site.
- The process **ends** with **negative feedback**, when the body detects the **blood clot** has been **formed**.

4) Positive feedback can also happen when a **homeostatic system breaks down**, e.g. if you're too cold for too long:

Hypothermia involves **positive feedback**:

- **Hypothermia** is **low body temperature** (below 35 °C).
- It happens when **heat's lost** from the body **quicker** than it can be **produced**.
- As body temperature **falls** the **brain doesn't work** properly and **shivering stops** — this makes body temperature **fall even more**.
- **Positive feedback** takes body temperature **further away** from the normal level, and it continues to decrease unless action is taken.

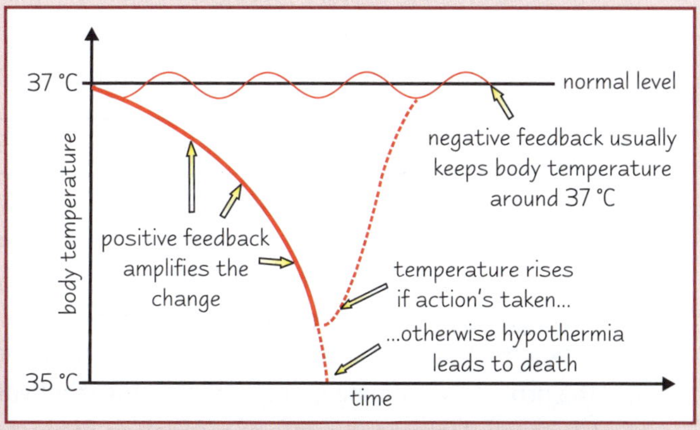

5) Positive feedback **isn't** involved in **homeostasis** because it **doesn't** keep your internal environment **stable**.

Warm-Up Questions

PRACTICE QUESTIONS

Q1 What is homeostasis and why is it necessary?

Q2 Why is it important to control blood pH?

Q3 Why is it important to control blood glucose concentration?

Exam Questions

Statement A: "Hyperthermia happens when the brain can't work properly and body temperature continues to increase."
Statement B: "When body temperature is low, mechanisms return the temperature to normal."

Q1 Look at statements A and B in the box.
   a) Which statement is describing a positive feedback mechanism? Give a reason for your answer. [1 mark]
   b) Describe and explain what effect a very high body temperature has on metabolic reactions. [2 marks]

Q2 Describe the importance of multiple negative feedback mechanisms in homeostasis. [2 marks]

Homeostasis works like a teacher — everything always gets corrected...

The key to understanding homeostasis is to get your head around negative feedback. Basically, if one thing goes up, the body responds to bring it down — and vice versa. When you're ready, turn over the page for some exciting examples.

Control of Blood Glucose Concentration

These pages are all about how homeostasis sorts out your blood glucose level so you can keep revising.

Eating and **Exercise** Change the **Concentration** of **Glucose** in your **Blood**

1) **All cells** need a constant **energy supply** to work — so **blood glucose concentration** must be carefully **controlled**.
2) The **concentration** of **glucose** in the blood is **normally** around **90 mg per 100 cm^3** of blood. It's **monitored** by cells in the **pancreas**.
3) Blood glucose concentration **rises** after **eating food** containing **carbohydrate**. Blood glucose concentration **falls** after **exercise**, as **more glucose** is used in **respiration** to **release energy**.

Insulin and **Glucagon Control** Blood Glucose Concentration

The hormonal system **controls** blood glucose concentration using **two hormones** called **insulin** and **glucagon**. Like all hormones, insulin and glucagon **travel** in the **blood** to their **target cells** (**effectors**). They're both **secreted** by clusters of cells in the **pancreas** called the **islets of Langerhans**:

- **Beta** (β) **cells** secrete **insulin** into the blood.
- **Alpha** (α) **cells** secrete **glucagon** into the blood.

Insulin and glucagon act on **effectors**, which respond to **restore** the blood glucose concentration to the **normal level**:

Insulin lowers blood glucose concentration when it's **too high**

1) Insulin binds to **specific receptors** on the cell membranes of **liver cells** and **muscle cells**.
2) It **increases** the **permeability** of muscle-cell membranes to glucose, so the cells **take up more glucose**. This involves **increasing** the number of **channel proteins** in the cell membranes (see next page for more).
3) Insulin also **activates enzymes** in liver and muscle cells that convert **glucose** into **glycogen**.
4) The cells are able to **store glycogen** in their cytoplasm, as an **energy source**.
5) The process of **forming glycogen** from glucose is called **glycogenesis**.
6) Insulin also **increases** the **rate** of **respiration** of glucose, especially in muscle cells.

Liver cells are also called hepatocytes.

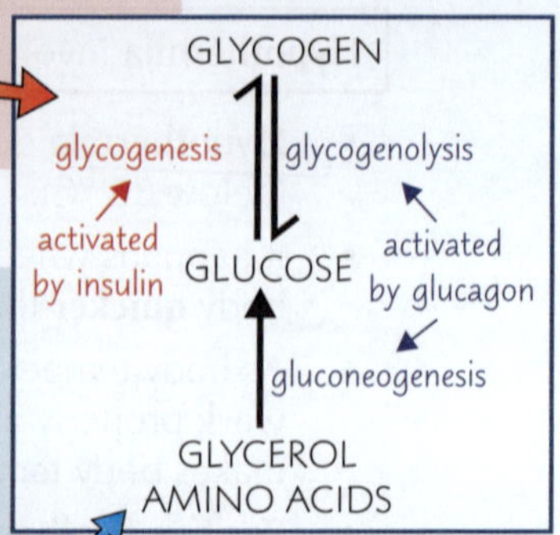

Glucagon raises blood glucose concentration when it's **too low**

1) Glucagon binds to **specific receptors** on the cell membranes of **liver cells**.
2) Glucagon **activates enzymes** in liver cells that **break down glycogen** into **glucose**.
3) The process of **breaking down glycogen** is called **glycogenolysis**.
4) Glucagon also activates **enzymes** that are involved in the formation of glucose from **glycerol** (a component of lipids) and **amino acids**.
5) The process of **forming glucose** from **non-carbohydrates** is called **gluconeogenesis**.
6) Glucagon **decreases** the **rate** of **respiration** of glucose in cells.

Because they **travel in the blood** to their target cells, the responses produced by **hormones** are **slower** than those produced by **nervous impulses** (which are very quick — see page 29). Hormones are not broken down as quickly as neurotransmitters though, so their effects tend to **last for longer**.

Negative Feedback Mechanisms Keep Blood Glucose Concentration **Normal**

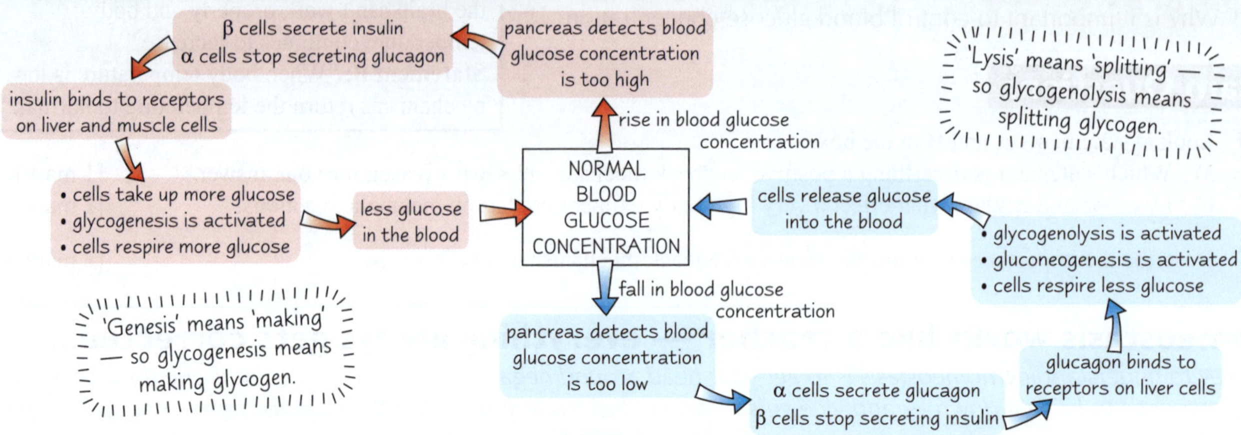

'Lysis' means 'splitting' — so glycogenolysis means splitting glycogen.

'Genesis' means 'making' — so glycogenesis means making glycogen.

Control of Blood Glucose Concentration

Insulin Makes **Glucose Transporters** Available for **Facilitated Diffusion**

1) **Skeletal** and **cardiac muscle cells** contain a **channel protein** called GLUT4. GLUT4 is a **glucose** transporter.
2) When **insulin** levels are **low**, GLUT4 is **stored** in **vesicles** in the **cytoplasm** of cells.
3) When insulin **binds** to **receptors** on the cell-surface membrane, it triggers the **movement** of GLUT4 **to the membrane**.
4) Glucose can then be transported **into the cell** through the GLUT4 protein, by **facilitated diffusion**.

Like **Glucagon, Adrenaline** Also **Increases Blood Glucose Concentration**

1) **Adrenaline** is a **hormone** that's secreted from your **adrenal glands** (found just above your kidneys).
2) It's secreted when there's a **low concentration** of **glucose** in your blood, when you're **stressed** and when you're **exercising**.
3) Adrenaline binds to **receptors** in the cell membrane of **liver cells**:
   - It **activates glycogenolysis** (the breakdown of glycogen to glucose).
   - It **inhibits glycogenesis** (the synthesis of glycogen from glucose).
4) It also **activates glucagon secretion** and **inhibits insulin secretion**, which increases glucose concentration.
5) Adrenaline gets the **body ready** for **action** by making **more glucose** available for **muscles** to respire.

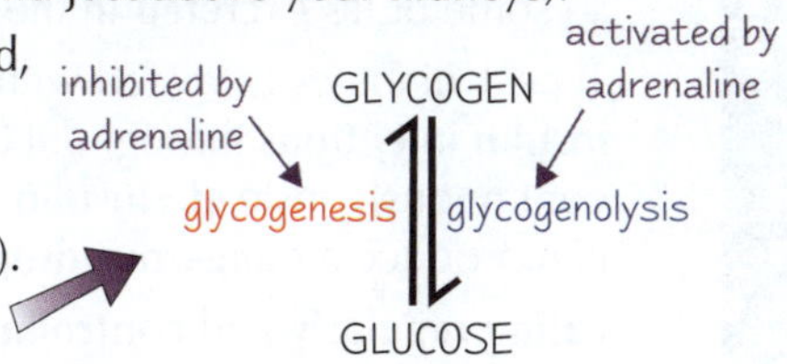

Adrenaline and **Glucagon** Act via a **Second Messenger**

Both **adrenaline** and **glucagon** can activate glycogenolysis **inside** a cell even though they bind to **receptors** on the **outside** of the cell. Here's **how** they do it:

- The **receptors** for adrenaline and glucagon have specific **tertiary structures** that make them complementary in shape to their respective hormones. Adrenaline and glucagon **bind** to their receptors and **activate** an **enzyme** called **adenylate cyclase** (also known as adenylyl cyclase).
- Activated adenylate cyclase converts **ATP** into a **chemical signal** called a '**second messenger**'.
- The second messenger is called **cyclic AMP** (**cAMP**).
- cAMP **activates** an enzyme called **protein kinase A**. Protein kinase A activates a **cascade** (a chain of reactions) that breaks down glycogen into glucose (**glycogenolysis**).

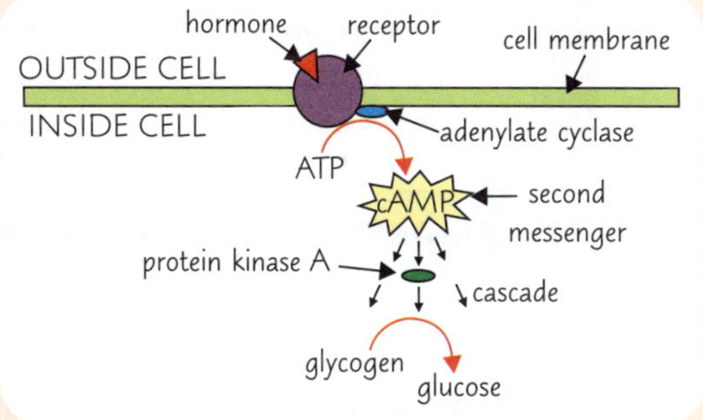

Warm-Up Questions

Q1 Why does your blood glucose concentration fall after exercise?

Q2 What's the process of breaking down glycogen into glucose called?

Q3 Give two effects of glucagon on liver cells.

PRACTICE QUESTIONS

Exam Questions

Q1 The pancreas secretes hormones that control blood glucose concentration.

a) What type of feedback mechanism is involved in the control of blood glucose concentration? Give a reason for your answer. [1 mark]

b) Describe the role of insulin in this feedback mechanism. [3 marks]

Q2 Glucagon and adrenaline trigger glycogenolysis when they bind to receptors on cell membranes. Explain how glycogenolysis is triggered inside the cell when these hormones bind to receptors on the cell surface. [3 marks]

My alpha cells detect low glucose — urgent tea and biscuit break needed...

Aaaaargh there are so many stupidly complex names to learn and they all look and sound exactly the same to me. You can't even get away with sneakily misspelling them all in your exam — like writing 'glycusogen' or 'gluconesisolysis'.

Control of Blood Glucose Concentration

Homeostasis doesn't always work. One example of this is diabetes...

Diabetes Occurs when **Blood Glucose Concentration** is **Not Controlled**

Diabetes mellitus is a condition where **blood glucose** concentration **can't** be **controlled** properly. There are **two types**:

Type I

1) In **Type I** diabetes, the immune system attacks the **β cells** in the islets of Langerhans so they **can't produce** any **insulin**. No one knows exactly what **causes** the immune system to do this. Scientists have found that some people have a **genetic predisposition** to developing Type I diabetes. They also think that the disease may be triggered by a **viral infection**.
2) After **eating**, the blood glucose level **rises** and **stays high** — this is called **hyperglycaemia** and can result in **death** if left untreated. The kidneys **can't reabsorb** all this glucose, so some of it's **excreted** in the urine.
3) Type 1 diabetes is treated with **insulin therapy**. Most people with Type I diabetes need **regular insulin injections** throughout the day, but some people use an **insulin pump** to deliver insulin continuously instead. Insulin therapy has to be **carefully controlled** because too much insulin can produce a **dangerous drop** in blood glucose levels — this is called **hypoglycaemia**.
4) **Eating regularly** and **controlling simple carbohydrate intake** (intake of sugars) helps to **avoid** a **sudden rise** in glucose.

Type II

1) **Type II** diabetes is usually acquired **later** in **life** than Type I. It is often linked with **obesity** and is **more likely** in people with a **family history** of the condition. Other risk factors include **lack of exercise**, **age** and **poor diet**.
2) It occurs when the β cells **don't produce enough insulin** or when the body's **cells don't respond** properly to **insulin**. Cells don't respond properly because the insulin **receptors** on their membranes **don't work** properly, so the cells **don't** take up enough glucose. This means the **blood glucose concentration** is **higher** than normal.
3) It can be treated by **eating a healthy**, **balanced diet**, **losing weight** (if necessary) and **regular exercise**. **Glucose-lowering medication** can be taken if diet and exercise can't control it. Eventually, **insulin injections** may be needed.

Type II Diabetes is a **Growing Health Problem**

1) **Type II diabetes** is becoming **increasingly common** in the UK. This has been linked to increasing levels of **obesity**, a move towards more **unhealthy diets** and **low levels of physical activity**.
2) Type II diabetes can cause **additional health problems**, including visual impairment and kidney failure, so **health advisors** are understandably keen to **educate people** about the **risks** and **reduce** the **incidence** of the disease. Some people also think the **food industry** has a role to play in tackling the problem.
3) You need to understand the various **responses** to the increase in Type II diabetes and be able to **evaluate** them.

To **reduce** the **risk** of developing Type II diabetes, **health advisors** recommend that people:

- eat a **diet** that's **low** in **fat**, **sugar** and **salt**, with plenty of **whole grains**, **fruit** and **vegetables**,
- take **regular exercise**,
- **lose weight** if necessary.

Campaigns like the NHS's '**Change4Life**', aim to **educate people** on how to have a **healthier diet** and **lifestyle**, and so reduce their risk of developing conditions like Type II diabetes.

Health advisors have also challenged the food industry to **reduce** the **advertising** of **junk food** (particularly to children), to **improve** the **nutritional value** of their products, and to use **clearer labelling** on products — allowing consumers to make **healthier choices** about what to buy.

In **response** to criticism, some **food companies** have attempted to make their products more **healthy**, e.g.

- using **sugar alternatives** to sweeten food/drinks,
- **reducing** the **sugar**, **fat** and **salt** content of products.

Some people believe that diet varieties are not as good for health as they are claimed to be, e.g. there is some evidence to suggest that artificial sweeteners are linked to weight gain.

However, there is **pressure** on companies to **increase profits**. They say that the industry will only respond fully in the **long term**, as **public perception** about healthy eating **changes**.

Control of Blood Glucose Concentration

Colorimetry is Used to Determine the **Concentration** of a **Glucose Solution**

Normally, the **concentration** of **glucose** in **urine** is very **low** — between **0** and **0.8 mM**. **Higher** concentrations than this may indicate **diabetes** (although a blood test would be needed to confirm it). You need to be able to **determine** the **concentration** of **glucose** in a **'urine' sample**, using **colorimetry**. Here's how:

1) **Quantitative Benedict's reagent** is **different** to normal Benedict's reagent. When heated with glucose, the **initial blue colour** is **lost**, but a **brick-red precipitate** is **not** produced.
2) You can use a **colorimeter** to measure the **light absorbance** of the solution **after** the quantitative Benedict's test has been carried out.
3) The **higher** the **concentration** of **glucose**, the **more blue colour** will be **lost** (i.e. the paler the solution will become), **decreasing** the **absorbance** of the solution.

Don't worry, it won't be real urine! You'll be given a fake sample by your teacher.

This is How You Do it:

Initially you need to make up several glucose solutions of **different**, **known concentrations**. You can do this using a **serial dilution** technique:

This is how you'd make **five serial dilutions** with a **dilution factor of 2**, starting with an initial glucose concentration of **4 mM**...

1) Line up five **test tubes** in a rack.
2) Add **10 cm³** of the initial **4 mM glucose solution** to the first test tube and **5 cm³ of distilled water** to the other four test tubes.
3) Then, using a pipette, draw **5 cm³** of the solution from the **first** test tube, add it to the distilled water in the **second** test tube and **mix** the solution **thoroughly**. You now have **10 cm³** of solution that's **half as concentrated** as the solution in the first test tube (it's **2 mM**).
4) Repeat this process **three more times** to create solutions of **1 mM**, **0.5 mM** and **0.25 mM**.

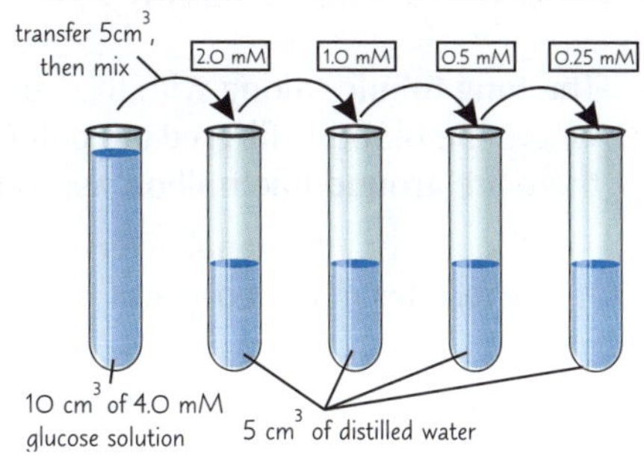

Once you've got your glucose solutions, you need to make a **calibration curve**. Here's how:

1) Do a **quantitative Benedict's test** on each solution (plus a **negative control** of **pure water**). Use the **same amount** of Benedict's solution in each case.
   To do the quantitative Benedict's test, you add quantitative Benedict's reagent to a sample and heat it in a water bath that's been brought to the boil.
2) Use a **colorimeter** (with a **red filter**) to measure the **absorbance** of the Benedict's solution **remaining** in each tube.
3) Use the results to make the **calibration curve**, showing absorbance against glucose concentration.

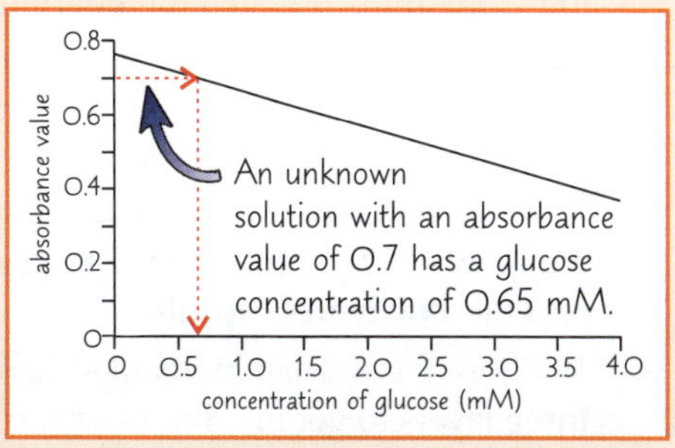

Then you can test the **unknown solution**, i.e. the **'urine' sample**, in the same way as the known concentrations, and use the calibration curve to find the concentration of glucose in the sample.

Warm-Up Questions

Q1 Briefly describe how you would produce a calibration curve to find the concentration of glucose in an unknown solution.

Exam Question

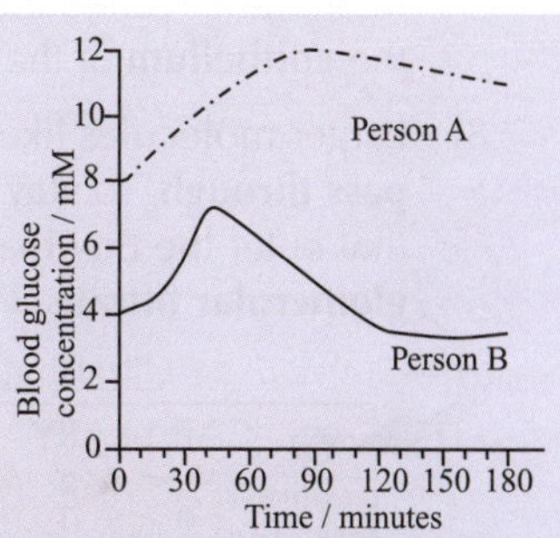

Q1 A glucose tolerance test is a medical test that can indicate the presence of diabetes. After fasting for 12 hours, a drink containing glucose is consumed. The graph shows how the blood glucose concentration of two people changed after having the drink. Person A has Type II diabetes. Person B does not have diabetes.

a) Give two pieces of evidence from the graph that suggest person A has diabetes. [2 marks]

b) Person A produces enough insulin but can't control their blood glucose concentration. Explain why. [2 marks]

Benedict's reagent makes you happy — it causes a loss of blues...

Evaluating can be tricky — you've got to give evidence to support your statements and look at both sides. You can't just say things like 'if only the food industry wasn't producing all that delicious junk food and making us all obese'.

The Kidneys

The kidneys make your urine by filtering waste products out of your blood and reabsorbing the useful stuff.

The **Kidneys Excrete Waste** and **Regulate Blood Water Potential**

1) One of the main functions of the kidneys is to **excrete waste products**, such as **urea**.
2) The kidneys also **regulate** the **water potential** of the **blood** — see pages 54-55.
3) As the **blood** passes through **capillaries** in the **cortex** (outer layer) of the **kidneys**, substances are **filtered out** of the blood and into **long tubules** that surround the capillaries. This process is called **ultrafiltration**.
4) Useful substances, such as **glucose** and the right amount of **water**, are then **reabsorbed** back into the **blood**. This process is called **selective reabsorption**.
5) The remaining **unwanted** substances pass along to the **bladder** and are excreted as **urine**.

Blood is **Filtered** at the **Start** of the **Nephrons**

The **long tubules** along with the bundle of **capillaries** where the blood is **filtered** are called **nephrons** — there are **around one million** nephrons in each kidney.

1) Blood from the **renal artery** enters smaller **arterioles** in the **cortex** of the kidney.
2) Each arteriole splits into a structure called a **glomerulus** (plural, glomeruli) — a **bundle** of **capillaries** looped inside a hollow ball called a **Bowman's capsule**.
3) This is where **ultrafiltration** takes place.
4) The **arteriole** that takes blood **into** each glomerulus is called the **afferent** arteriole, and the arteriole that takes the filtered blood **away** from the glomerulus is called the **efferent** arteriole.
5) The **efferent** arteriole is **smaller** in **diameter** than the afferent arteriole, so the blood in the glomerulus is under **high pressure**.
6) The high pressure **forces liquid** and **small molecules** in the blood **out** of the **capillary** and **into** the **Bowman's capsule**.
7) The liquid and small molecules pass through **three** layers to get into the Bowman's capsule and **enter** the nephron **tubules** — the **capillary wall**, a membrane (called the **basement membrane**) and the **epithelium** of the Bowman's capsule.
8) Larger molecules like **proteins** and **blood cells can't pass through**, so **stay** in the blood. The substances that enter the Bowman's capsule are known as the **glomerular filtrate**.

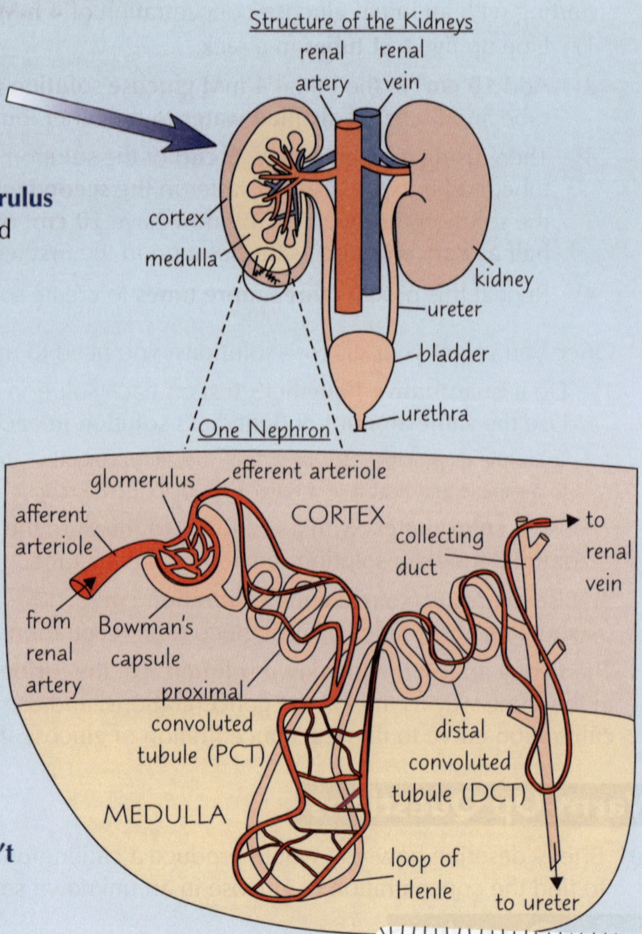

You need to learn the structure of a nephron.

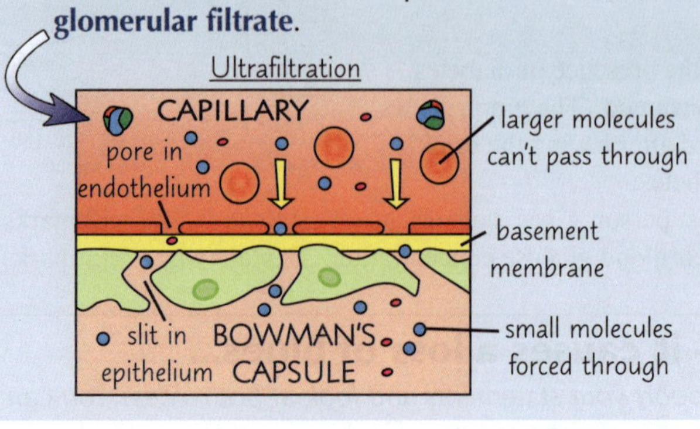

9) The **glomerular filtrate** passes along the rest of the nephron and **useful substances** are **reabsorbed** along the way — see next page.
10) Finally, the filtrate flows through the **collecting duct** and passes out of the kidney along the **ureter**.

The Kidneys

Useful Substances are Reabsorbed Along the Nephron Tubules

1) **Selective reabsorption** takes place as the glomerular filtrate flows along the **proximal convoluted tubule** (**PCT**), through the **loop of Henle**, and along the **distal convoluted tubule** (**DCT**).
2) **Useful substances** leave the tubules of the nephrons and **enter** the capillary network that's **wrapped** around them (see diagram on previous page).
3) The **epithelium** of the wall of the PCT has **microvilli** to provide a **large surface area** for the **reabsorption** of useful materials from the **glomerular filtrate** (in the tubules) into the **blood** (in the capillaries).
4) Useful solutes, like **glucose**, are reabsorbed along the PCT by **active transport** and **facilitated diffusion**.
5) **Water** enters the blood by **osmosis** because the **water potential** of the blood is **lower** than that of the filtrate. Water is reabsorbed from the **PCT**, **loop of Henle**, **DCT** and the **collecting duct** (see next page).
6) The filtrate that remains is **urine**, which passes along the **ureter** to the **bladder**.

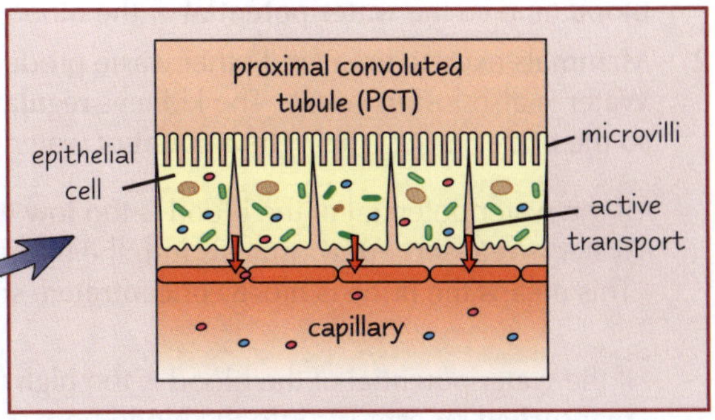

Remember: water potential describes the tendency of water to move from one area to another. Water will move from an area of higher water potential to an area of lower water potential — it moves down the water potential gradient.

Urine is usually **made up of**:
- **Water** and **dissolved salts**.
- **Urea**.
- Other substances such as **hormones** and **excess vitamins**.

Urine **doesn't** usually contain:
- **Proteins** and **blood cells** — they're **too big** to be **filtered out** of the blood.
- **Glucose** because it's **actively reabsorbed** back into the blood (see above).

The volume of water in urine varies depending on how much you've drunk (see p. 55).

Warm-Up Questions

Q1 What is selective reabsorption?

Q2 Which blood vessel supplies the kidney with blood?

Q3 What are the bundles of capillaries found in the cortex of the kidneys called?

Q4 By which two processes is glucose reabsorbed from the PCT?

Exam Question

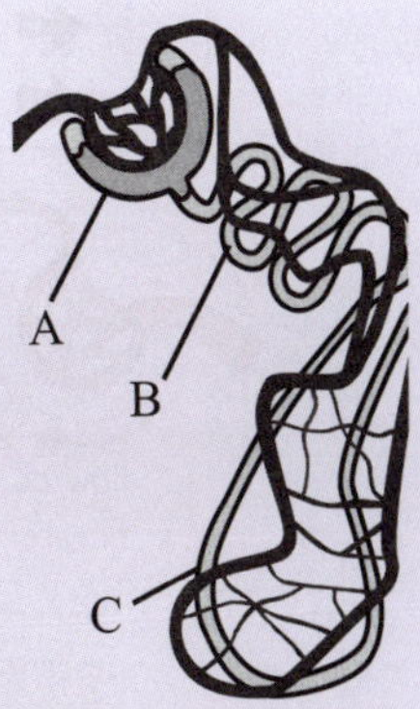

Q1 The diagram on the right shows part of a nephron.

a) Explain how glomerular filtrate is formed at point A. [2 marks]

b) Would you expect the concentration of glucose to be lower at point B or point C on the diagram? Explain your answer. [1 mark]

c) The rate at which the kidneys filter blood is called the glomerular filtration rate (GFR). GFR is normally around 6300 cm^3 $hour^{-1}$. Tests revealed 0 mg of glucose in a person's urine. The same person's blood glucose concentration was 0.9 mg cm^{-3}. Assuming a normal GFR, calculate the rate at which glucose is reabsorbed back into the blood. Give your answer in mg min^{-1}. [1 mark]

Mmm — it's steak and excretion organ pie for dinner...

Excretion is a pretty horrible sounding word I know, but it's gotta be done. Mind you, I've never been able to eat kidney ever since I learnt all about this urine production business. Shame really — I used to love kidney sarnies for lunch. Make sure you can describe how the glomerular filtrate is formed and how glucose and water are reabsorbed.

Controlling Blood Water Potential

The kidneys control the water potential of the blood — osmoregulation, if you're being posh.

The **Kidneys** Regulate the **Water Potential** of the **Blood**

1) Water is **essential** to keep the body **functioning**, so the **amount** of water in the **blood** (and so the **water potential** of the blood) needs to be kept **constant**.
2) Mammals excrete **urea** (and other waste products) in **solution**, which means **water** is **lost** during excretion. Water is also lost in **sweat**. The kidneys **regulate** the **water potential** of the blood (and urine), so the body has just the **right amount** of water — this is called **osmoregulation**.

If the water potential of the blood is too **low** (the body is **dehydrated**), **more** water is **reabsorbed** by osmosis **into** the blood from the tubules of the nephrons. This means the urine is **more concentrated**, so **less** water is **lost** during excretion.

If the water potential of the blood is too **high** (the body is too **hydrated**), **less** water is **reabsorbed** by osmosis **into** the blood from the tubules of the nephrons. This means the urine is **more dilute**, so **more** water is **lost** during excretion (see next page).

Brad liked his urine to be dilute.

3) Water is **reabsorbed** into the blood along **almost all** of the **nephron** (see previous page), but **regulation of water potential** mainly takes place in the **loop of Henle**, **DCT** and **collecting duct**. The **volume** of water reabsorbed by the DCT and collecting duct is controlled by **hormones** (see next page).

The **Loop of Henle** Maintains a **Sodium Ion Gradient**

The **loop of Henle** is located in the **medulla** (inner layer) of the kidneys. It's made up of two **'limbs'** — the **descending** limb and the **ascending** limb. The limbs control the movement of **sodium ions** so that water can be **reabsorbed** by the blood.

(1) Near the **top** of the **ascending** limb, **Na^+** ions are **pumped out** into the **medulla** using **active transport**. The ascending limb is **impermeable** to **water**, so the water **stays inside** the tubule. This creates a **low water potential** in the **medulla**, because there's a **high concentration** of ions.

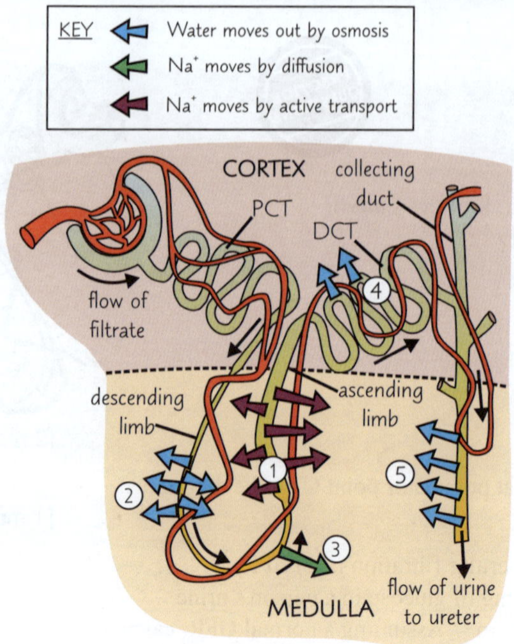

(2) Because there's a **lower** water potential in the **medulla** than in the descending limb, **water** moves **out** of the **descending limb** (which is permeable to water) **into** the **medulla** by **osmosis**. This makes the **filtrate more concentrated** (the ions can't diffuse out — the descending limb isn't permeable to them). The water in the medulla is **reabsorbed** into the **blood** through the **capillary network**.

(3) Near the **bottom** of the **ascending** limb Na^+ ions **diffuse out** into the **medulla**, further **lowering** the **water potential** in the medulla. The ascending limb is **impermeable** to **water**, so it **stays in the tubule**.

(4) Water moves out of the **distal convoluted tubule** (DCT) by osmosis and is reabsorbed into the blood.

(5) The first three stages massively **increase** the **ion concentration** in the **medulla**, which **lowers** the **water potential**. This causes **water** to **move out** of the **collecting duct** by **osmosis**. As before, the water in the medulla is **reabsorbed** into the **blood** through the **capillary network**.

The **volume** of water **reabsorbed** into the capillaries is **controlled** by **changing the permeability** of the **DCT** and the **collecting duct** (see next page).

Controlling Blood Water Potential

Water Reabsorption is Controlled by Hormones

1) The water potential of the blood is **monitored** by cells called **osmoreceptors** in a part of the **brain** called the **hypothalamus**.
2) When the water potential of the blood **decreases**, water will move **out** of the osmoreceptor cells by **osmosis**. This causes the cells to **decrease** in **volume**. This sends a signal to other cells in the **hypothalamus**, which send a signal to the **posterior pituitary gland**. This causes the posterior pituitary to release a **hormone** called **antidiuretic hormone** (ADH) into the blood.
3) ADH makes the walls of the DCT and collecting duct **more permeable** to **water**.
4) This means **more water** is **reabsorbed** from these tubules **into** the medulla and into the blood by osmosis. A **small** amount of **concentrated urine** is produced, which means **less water** is **lost** from the body.

It's called antidiuretic hormone because diuresis is when lots of dilute urine is produced, so anti means a small amount of concentrated urine is produced.

Here's how ADH changes the **water content** of the **blood** when it's too **low** or too **high**:

1 Blood ADH Level Rises When You're Dehydrated

Dehydration is what happens when you **lose water**, e.g. by sweating during exercise, so the **water content** of the blood needs to be **increased**:

1) The **water content** of the blood **drops**, so its **water potential drops**.
2) This is detected by **osmoreceptors** in the **hypothalamus**.
3) The **posterior pituitary gland** is stimulated to release **more ADH** into the blood.
4) **More ADH** means that the DCT and collecting duct become **more permeable**, so **more water** is **reabsorbed** into the blood by osmosis.
5) A **small amount** of **highly concentrated** urine is produced and **less water** is **lost**.

Dehydrated? Me? As if...

2 Blood ADH Level Falls When You're Hydrated

If you're **hydrated**, you've taken in **lots of water**, so the **water content** of the blood needs to be **reduced**:

1) The **water content** of the blood **rises**, so its **water potential rises**.
2) This is detected by the **osmoreceptors** in the **hypothalamus**.
3) The **posterior pituitary gland** releases **less ADH** into the blood.
4) **Less ADH** means that the DCT and collecting duct become **less permeable**, so **less water** is **reabsorbed** into the blood by osmosis.
5) A **large amount** of **dilute** urine is produced and **more water** is **lost**.

Warm-Up Questions

Q1 Describe what happens along the descending limb of the loop of Henle.

Q2 Which cells monitor the water content of the blood?

PRACTICE QUESTIONS

Exam Questions

Q1 The level of ADH in the blood rises during strenuous exercise.

a) Explain the cause of the increase in ADH. [4 marks]

b) Explain the effect that the increased ADH levels have on kidney function. [2 marks]

Q2 Gerbils have longer loops of Henle than mice.
Suggest and explain how this helps gerbils to produce less urine than mice. [4 marks]

If you don't understand what ADH does, ur-ine trouble...

There are two main things to learn here — how a sodium ion gradient lets the kidneys reabsorb so much water into the blood and how the water content of the blood is regulated by osmoreceptors in the hypothalamus. Now I need a wee.

Extra Exam Practice

Before you tick Topic 6 off your list, have a go at these questions mixing different parts of the section.

- Have a look at this example of how to answer a tricky exam question.
- Then check how much you've understood from Topic 6 by having a go at the questions on the next page.

1 The autonomic nervous system controls heart rate in response to internal stimuli.

1.1 Pain can increase heart rate. Pain can be felt following stimulation of receptors called nociceptors. Stimulation of a nociceptor increases the permeability of its membrane to ions. Suggest an explanation for how stimulation of nociceptors leads to the brain rapidly sensing pain.

(3 marks)

Changes in heart rate affect the volume of blood that is pumped around the body in a given time. **Figure 1** shows the volume of blood in the left ventricle of a person over time.

Figure 1

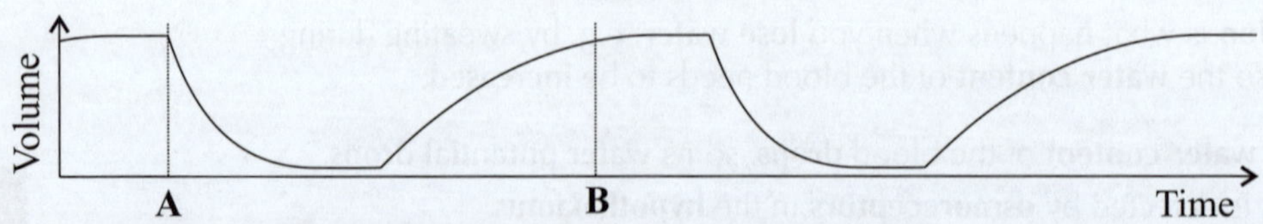

1.2 With reference to the electrical activity of the heart, explain the change in the volume of blood in the ventricle between points **A** and **B** in **Figure 1**.

(4 marks)

1.3 The action of acetylcholine in the heart reduces the rate at which cells in the SAN depolarise. Atropine is a drug that blocks cholinergic receptors on the SAN. Suggest how atropine causes an increase in heart rate.

(3 marks)

1.1

The increased permeability of the membrane to ions will alter the **potential difference** of the membrane. This will lead to a **generator potential** big enough to trigger an **action potential**. Action potentials will travel along sensory neurones to the brain, which will interpret the signals as feelings of pain.

You're not expected to know anything specific about nociceptors — you just have to apply what you know about what happens when a receptor is stimulated.

1.2

At **point A**, a wave of electrical activity is passing through the **Purkyne tissue**. This causes the ventricle to contract and push blood out of the heart, until the volume drops to the **lowest point**. The **volume** then **plateaus** until the **SAN** sends out another wave of electrical activity that stimulates the atrium to contract, pushing blood back into the ventricle and increasing the volume of blood it contains, until it reaches **point B**.

Don't get carried away writing more than you need to — you've been asked to describe the changes between points A and B so only mention this section of the graph.

You'll need to mention these specific structures in the heart to pick up marks.

1.3

Atropine will prevent acetylcholine from affecting the SAN, as it blocks the **cholinergic receptors** that **acetylcholine acts upon**. Therefore, neurotransmitters from the **sympathetic branch** of the autonomic nervous system will have **more of an effect** on the rate at which the SAN depolarises, so **heart rate** will **increase**.

Always double-check you're answering the question — here you've got to explain why heart rate might increase (not just why it might be prevented from decreasing).

From your knowledge of cholinergic synapses you should know that acetylcholine works at cholinergic receptors.

Extra Exam Practice

2 It is estimated that over 4 million people in the UK have diabetes.

Diabetic retinopathy is an eye condition associated with diabetes. It can cause blurred vision by damaging the retina. The extent of vision impairment partly depends on which areas of the retina are most damaged, as different areas of the retina have different proportions of cones and rods.

2.1 With reference to the connections cones and rods make to the optic nerve, explain why vision may be more blurred if diabetic retinopathy affects areas of the retina with a greater proportion of cones than rods.

(3 marks)

SGLT2 inhibitors are drugs that are used to treat type 2 diabetes. They block glucose transport proteins in the proximal convoluted tubules of the kidneys.

2.2 Explain how SGLT2 inhibitors will affect the glucose concentration in the urine.

(2 marks)

Colorimetry and a glucose calibration curve were used to determine the glucose concentration in the urine of a patient taking SGLT2 inhibitors. **Figure 2** shows the glucose calibration curve used.

Figure 2

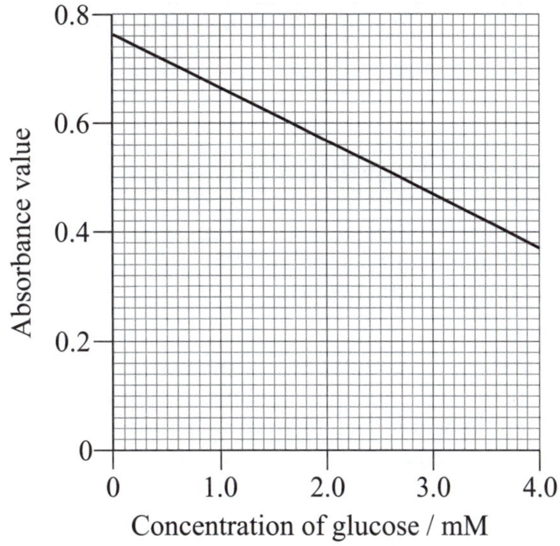

2.3 The absorbance value of the urine was 0.5. The colorimeter had an uncertainty of ±0.05. Using **Figure 2** and the uncertainty, determine the range in which the glucose concentration of this sample may lie.

(2 marks)

3 A scientist investigated muscle contraction during exercise.

3.1 Explain how the phosphocreatine stored in a cell can be used during muscle contraction.

(2 marks)

3.2 Explain how the movement of calcium ions lengthens the sarcomere following muscle contraction.

(3 marks)

3.3 Repeated muscle contraction during exercise can lead to a reduction in blood water potential. Diabetes insipidus is a condition in which a person secretes lower levels of ADH than normal. Would a person with diabetes insipidus be more or less at risk of dehydration during exercise than a person without the condition? Explain your answer.

(4 marks)

Inheritance

Nope, this isn't about who gets Mum's best china — we're talking genetic inheritance here...

You **Need to Know** These **Genetic Terms**

'Codes for' means 'contains the instructions for'.

TERM	DESCRIPTION
Gene	A sequence of bases on a DNA molecule that codes for a protein (polypeptide), which results in a characteristic, e.g. a gene for eye colour.
Allele	A different version of a gene. There can be many different alleles of a single gene, but most plants and animals, including humans, only carry two alleles of each gene, one from each parent. The order of bases in each allele is slightly different — they code for different versions of the same characteristic. They're represented using letters, e.g. the allele for brown eyes (B) and the allele for blue eyes (b).
Genotype	The genetic constitution of an organism — the alleles an organism has, e.g. BB, Bb or bb for eye colour.
Phenotype	The expression of the genetic constitution and its interaction with the environment — an organism's characteristics, e.g. brown eyes.
Dominant	An allele whose characteristic appears in the phenotype even when there's only one copy. Dominant alleles are shown by a capital letter. E.g. the allele for brown eyes (B) is dominant — if a person's genotype is Bb or BB, they'll have brown eyes.
Recessive	An allele whose characteristic only appears in the phenotype if two copies are present. Recessive alleles are shown by a lower case letter. E.g. the allele for blue eyes (b) is recessive — if a person's genotype is bb, they'll have blue eyes.
Codominant	Alleles that are both expressed in the phenotype — neither one is recessive, e.g. the alleles for haemoglobin.
Locus	The fixed position of a gene on a chromosome. Alleles of a gene are found at the same locus on each chromosome in a pair.
Homozygote	An organism that carries two copies of the same allele, e.g. BB or bb.
Heterozygote	An organism that carries two different alleles, e.g. Bb.
Carrier	A person carrying an allele which is not expressed in the phenotype but that can be passed on to offspring.

Genetic Diagrams Show the **Possible Genotypes** of **Offspring**

Humans are **diploid** organisms (we have two sets of chromosomes) so we have **two alleles** for **each gene**. **Gametes** (sex cells) contain only **one allele** for each gene. When gametes from two parents fuse together, the alleles they contain form the **genotype** of the **offspring** produced. At each **locus**, the genotype can be **homozygous** or **heterozygous**.

Genetic diagrams can be used to **predict** the **genotypes** and **phenotypes** of the offspring produced if two parents are **crossed** (bred). You need to know how to use genetic diagrams to interpret or predict the results of various crosses, including **monohybrid crosses**. **Monohybrid inheritance** is the inheritance of a **characteristic** controlled by a **single gene**. **Monohybrid crosses** show the **likelihood** of the **different alleles** of that gene (and so different versions of the characteristic) being **inherited** by offspring of certain parents. This genetic diagram shows how **wing length** is inherited in fruit flies:

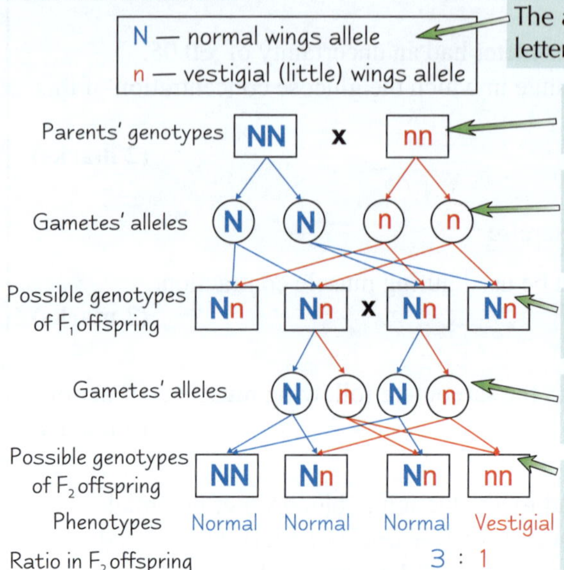

The allele for **normal wings** is **dominant**, so it's shown by a **capital** letter N. Any flies that have even one N allele will have normal wings.

One parent is **homozygous** with **normal wings** (**NN**) and one is **homozygous** with **vestigial wings** (**nn**).

The normal winged parent **only** produces gametes with the allele for **normal wings** (**N**). The vestigial winged parent **only** produces gametes with the allele for **vestigial wings** (**n**).

All F_1 offspring are **heterozygous** (**Nn**), as one allele is inherited from **each** parent.

The first set of offspring is called the F_1 generation.

The gametes produced by the F_1 offspring may contain the allele for **either normal** (**N**) or **vestigial wings** (**n**).

The F_2 offspring could have **either** normal or vestigial wings. But there's a **75%** chance they'll have the **normal wings phenotype** (genotype of **NN** or **Nn**) and a **25%** chance they'll have the **vestigial wings phenotype** (genotype **nn**). So you'd expect a **3:1** ratio of normal : vestigial wings in the offspring. **Whenever** you do a monohybrid cross with **two heterozygous** parents you get a **3:1** ratio of **dominant : recessive** characteristic.

The second set of offspring is called the F_2 generation.

Inheritance

A **Punnett square** is just another way of showing a **genetic diagram** — they're also used to predict the **genotypes** and **phenotypes** of offspring. The Punnett squares below show the same crosses from the previous page:

1) First work out the alleles the **gametes** would have.

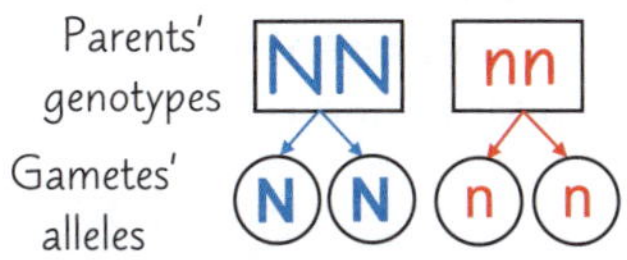

2) Next **cross** the **parents' gametes** to show the possible genotypes of the F_1 **generation** — **all heterozygous**, Nn.

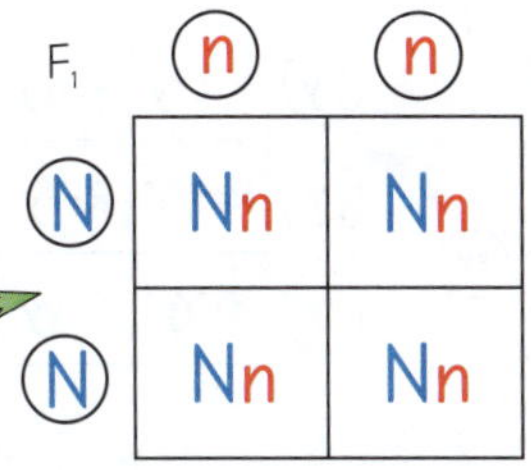

3) Then **cross** the **gametes' alleles** of the F_1 **generation** to show the possible **genotypes** of the F_2 **generation**. The Punnett square shows a **75%** chance that offspring will have **normal wings** and a **25%** chance that they'll have **vestigial wings**, i.e. a **3:1 ratio**.

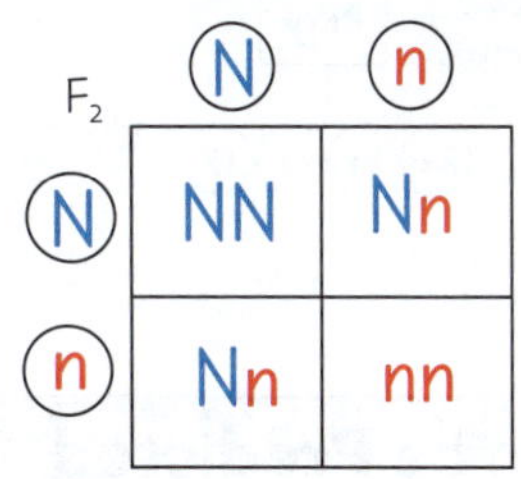

- 1 in 4 chance of offspring having the genotype NN (normal wings)
- 2 in 4 chance of offspring having the genotype Nn (normal wings)
- 1 in 4 chance of offspring having the genotype nn (vestigial wings)
- So, phenotype ratio normal:vestigial = 3:1

Some **Genes** Have **Codominant Alleles**

Occasionally, alleles show **codominance** — **both alleles** are expressed in the **phenotype**, **neither one** is recessive. One example in humans is the allele for **sickle-cell anaemia**:

1) People who are **homozygous** for **normal haemoglobin** (**H^NH^N**) don't have the disease.
2) People who are **homozygous** for **sickle haemoglobin** (**H^SH^S**) have **sickle-cell anaemia** — all their **blood cells** are **sickle-shaped** (crescent-shaped).
3) People who are **heterozygous** (**H^NH^S**) have an **in-between** phenotype, called the **sickle-cell trait** — they have **some** normal haemoglobin and some sickle haemoglobin. The two alleles are **codominant** because they're **both** expressed in the **phenotype**.
4) The **genetic diagram** on the right shows the possible offspring from **crossing** two parents with **sickle-cell trait** (**heterozygous**).

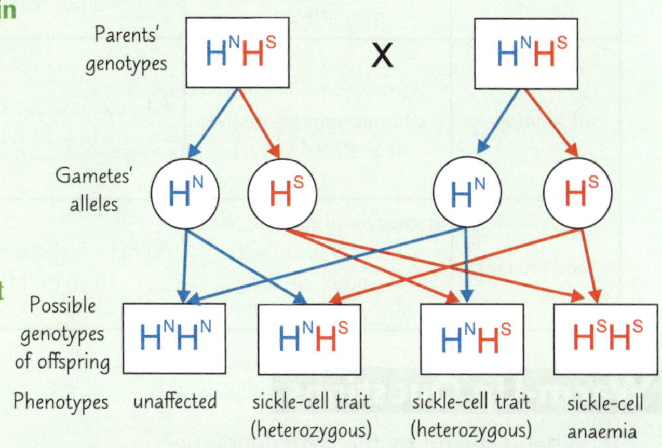

Some **Genes** Have **Multiple Alleles**

Inheritance is **more complicated** when there are **more than two** alleles of the same gene (**multiple alleles**).

Example In the **ABO blood group system** in humans there are **three alleles** for blood type:

I^O is the allele for blood group **O**. **I^A** is the allele for blood group **A**. **I^B** is the allele for blood group **B**.

Allele **I^O** is **recessive**. Alleles **I^A** and **I^B** are **codominant** — people with genotype **I^AI^B** will have blood group **AB**.

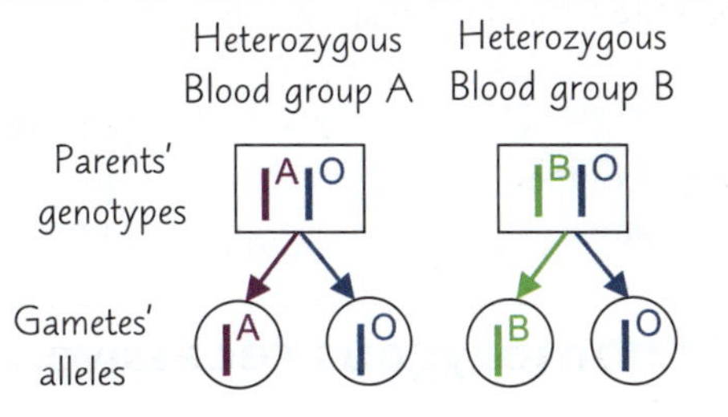

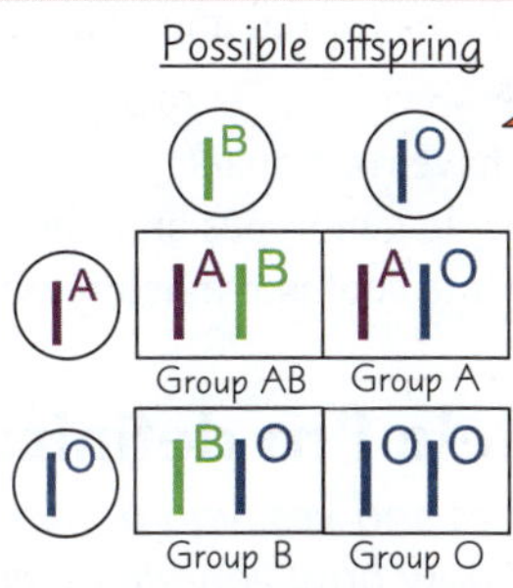

The genetic diagram shows a cross between a **heterozygous** person with blood group **A** and a **heterozygous** person with blood group **B**. Any offspring could have one of **four** different blood groups — **A**, **B**, **O** or **AB**.

Recessive blood groups are normally really rare, but it just so happens that loads of people in Britain are descended from people who were $I^O I^O$, so O's really common.

Inheritance

Genetic Diagrams can Show how More Than One Characteristic is Inherited

You can use genetic diagrams to work out the chances of offspring inheriting certain **combinations** of characteristics. For example, you can use a **dihybrid cross** to look at how **two different genes** are inherited at the same time. The diagram below is a **dihybrid cross** showing how seed texture **and** colour are inherited in **pea plants**.

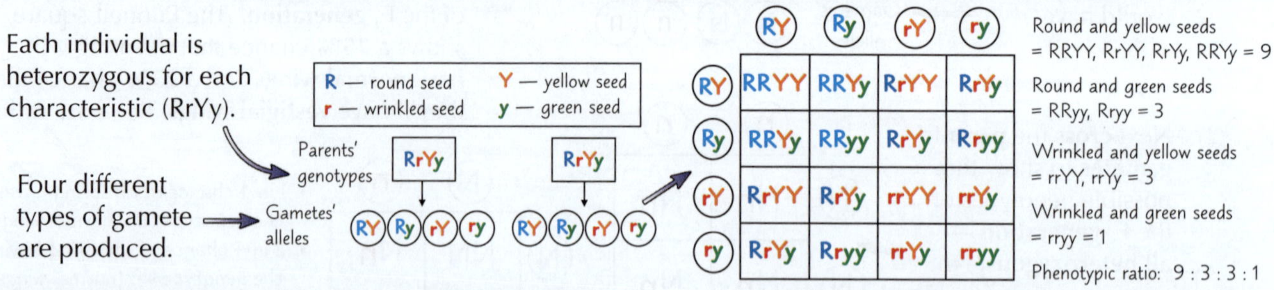

You can also do dihybrid crosses with codominant alleles. They work in the same way as this one but the phenotypic ratios produced are different.

Phenotypic Ratios can be Predicted

The **phenotypic ratio** is the **ratio** of **different phenotypes** in offspring.
Genetic diagrams allow you to **predict** the phenotypic ratios in F_1 **and** F_2 **offspring**.
Here's a handy summary table of ratios for the following crosses:

Type of Cross	Parents	Phenotypic Ratio in F_1	Phenotypic Ratio in F_2
Monohybrid	homozygous dominant x homozygous recessive (e.g. RR x rr)	All heterozygous offspring (e.g. Rr)	3 : 1 dominant : recessive
Dihybrid	homozygous dominant x homozygous recessive (e.g. RRYY x rryy)	All heterozygous offspring (e.g. RrYy)	9 : 3 : 3 : 1 dominant both : dominant 1st recessive 2nd : recessive 1st dominant 2nd : recessive both
Codominant	homozygous for one allele x homozygous for the other allele (e.g. H^NH^N x H^SH^S)	All heterozygous offspring (e.g. H^NH^S)	1 : 2 : 1 homozygous for one allele : heterozygous for the other allele

Sometimes you **won't** get the **expected** (predicted) phenotypic ratio — it'll be quite different. This can be because of **sex linkage**, **autosomal linkage** or **epistasis** — all of which are covered on pages 61-63.

Warm-Up Questions

Q1 What is meant by the term genotype?

Q2 What is meant by the term phenotype?

Q3 What does a dihybrid cross show you?

Exam Questions

Q1 In pea plants, seed texture (round or wrinkled) is passed from parent to offspring by monohybrid inheritance. The allele for round seeds is represented by R and the allele for wrinkled seeds is represented by r.

Draw a genetic diagram to show the possible genotypes of F_1 offspring produced by crossing a homozygous round-seed pea plant with a homozygous wrinkled seed pea plant. [3 marks]

Q2 Individuals of a particular breed of cow can have a red, white or roan coat. Animals with a roan coat have patches of both red and white hair. The alleles for red and white coats are C^R and C^W respectively. Heterozygotes for these alleles have roan coats.

a) Explain why heterozygotes for C^R and C^W have roan coats. [1 mark]

b) Draw a genetic diagram to predict the possible genotypes and phenotypes of the F_1 offspring produced by a parent with a white coat and a heterozygous parent. [4 marks]

If there's a dominant revision allele I'm definitely homozygous recessive...

OK, so there are a lot of fancy words on these pages and yes, you do need to know them all. Sorry. But don't despair — once you've learnt what the words mean and know how genetic diagrams work it'll all just fall into place.

Linkage and Epistasis

Right, this stuff is fairly hard, so if you don't get it first time don't panic just work through it again until you do...

Some **Characteristics** are **Sex-linked**

1) The genetic information for **gender** is carried on two **sex chromosomes**.
2) In mammals, **females** have **two X** chromosomes (XX) and **males** have **one X** and **one Y** chromosome (XY).
3) A **characteristic** is said to be **sex-linked** when the allele that codes for it is located on a **sex chromosome**.
4) The **Y chromosome** is **smaller** than the X chromosome and carries **fewer genes**. So most genes on the sex chromosomes are **only carried** on the X chromosome (called **X-linked** genes).
5) As **males** only have **one X chromosome**, they often only have **one allele** for sex-linked genes. So because they **only** have one copy, they **express** the **characteristic** of this allele even if it's **recessive**. This makes males **more likely** than females to show **recessive phenotypes** for genes that are sex-linked.
6) Genetic disorders caused by **faulty alleles** on sex chromosomes include **colour blindness** and **haemophilia**. The faulty alleles for both of these disorders are carried on the X chromosome — they're called **X-linked disorders**.

Example

1) **Colour blindness** is a **sex-linked disorder** caused by a faulty allele carried on the **X** chromosome.
2) As it's sex-linked **both** the chromosome and the allele are **represented** in the **genetic diagram**, e.g. X^n, where **X** represents the **X chromosome** and **n** the **faulty allele** for **colour vision**.
3) The **Y chromosome** doesn't have an allele for colour vision so is **just** represented by **Y**.
4) **Females** would need **two copies** of the **recessive allele** to be colour blind, while **males** only need **one copy**. This means colour blindness is **much rarer** in **women** than **men**.

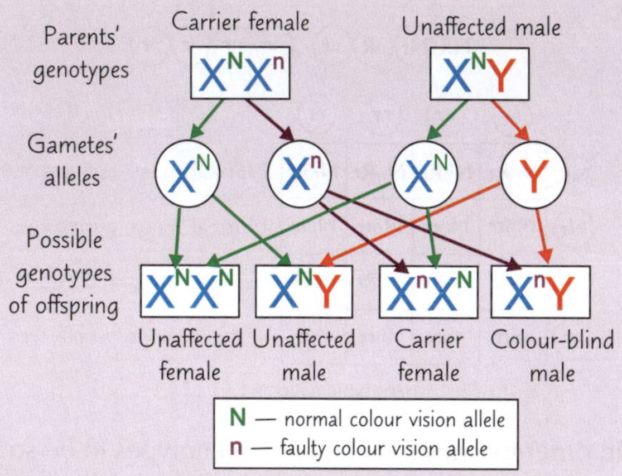

7) In the example above, there's a **3 : 1 ratio** of offspring **without** colour blindness : offspring **with** colour-blindness. But when a **female carrier** and a **male without colour-blindness** have children (as in this example), only their **male offspring** are at risk of being colour-blind. So you can also say that there's a predicted **2 : 1 : 1 ratio** — of **female** offspring **without** colour-blindness : **male** offspring **without** colour-blindness : **male** offspring **with** colour-blindness.
8) This ratio will **change** if a female carrier (X^NX^n) and a male **with** colour-blindness (X^nY) have children. The predicted ratio will then be **1 : 1** — of offspring **with** colour-blindness : offspring **without** colour-blindness. The ratio will be the **same** for offspring of **each gender**. You only end up with this predicted ratio for a monohybrid F_2 cross with a **sex-linked characteristic**.

Some **Autosomal** Genes are **Linked**

Crossing over is when two homologous (paired) chromosomes 'swap bits'. It happens in meiosis I before independent segregation. You'll have learnt about this in Year 1 of your course.

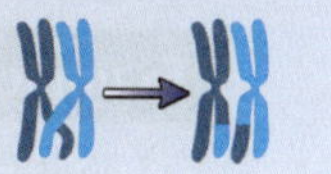

1) **Autosome** is the fancy name for any chromosome that **isn't** a sex chromosome. **Autosomal genes** are the genes located on the autosomes.
2) Genes on the **same autosome** are said to be **linked** — because they're on the same autosome they'll stay together during the **independent segregation of chromosomes** in meiosis I, and their alleles will be **passed on to the offspring together**. The only reason this won't happen is if **crossing over** splits them up first.
3) The **closer together** two genes are on the autosome, the **more closely** they are said to be **linked**. This is because **crossing over** is **less likely** to split them up.
4) If two genes are autosomally linked, you **won't get** the phenotypic ratio you expect in the offspring of a cross.

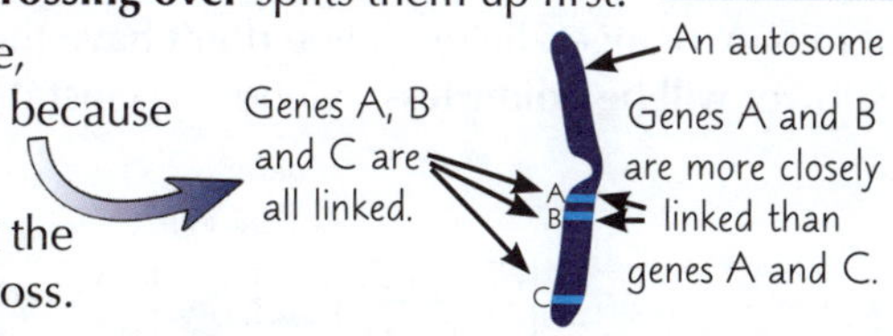

5) For example, in a **dihybrid cross** between two heterozygous parents you'd expect a **9 : 3 : 3 : 1 ratio** in the offspring (see previous page). Instead, the phenotypic ratio is more likely to be that expected for a **monohybrid cross** between two heterozygous parents (**3 : 1**) because the two autosomally-linked alleles are **inherited together**. This means that a **higher proportion** of the **offspring** will have their **parents'** (heterozygous) **genotype** and **phenotype**.
6) This allows you to use the **predicted phenotypic ratio** to **identify** autosomal linkage.

Linkage and Epistasis

Genetic Cross Results Can Show Autosomal Linkage

In the exam you might get some **genetic cross results** that show **linkage** and have to explain them.

Example

A scientist was investigating **linkage** between the **genes** for **eye colour** and **wing length** in **fruit flies**.

The gene for **normal wings** (**N**) is dominant to the gene for **vestigial wings** (**n**) and the gene for **red eyes** (**R**) is dominant to the gene for **purple eyes** (**r**). The **first cross** the scientist carried out was between flies **homozygous dominant** for both normal wings and red eyes (**NNRR**) and flies **homozygous recessive** for both vestigial wings and purple eyes (**nnrr**). The resulting offspring were all **heterozygous** for normal wings and red eyes (**NnRr**).

The **second cross** the scientist carried out was between these offspring (**NnRr**) and the flies homozygous recessive for vestigial wings and purple eyes (**nnrr**).

This is known as a back cross (crossing the offspring with one of the parents).

He expected a **1 : 1 : 1 : 1** ratio as shown below:

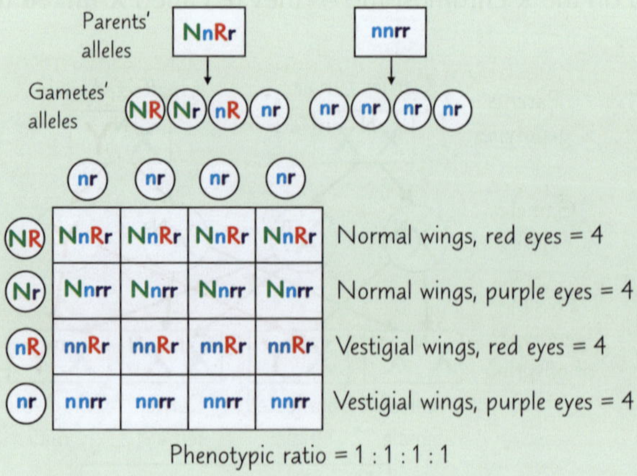

However, the results he got for this cross show a **8 : 1 : 1 : 8** ratio as in the **table**:

	Number of offspring
Normal wings, red eyes (NnRr)	1216
Normal wings, purple eyes (Nnrr)	152
Vestigial wings, red eyes (nnRr)	148
Vestigial wings, purple eyes (nnrr)	1184

Phenotypic ratio = 8 : 1 : 1 : 8

In order for the **NnRr** and **nnrr** genotypes to be so **common** in the offspring, the **NR** alleles and the **nr** alleles in the **NnRr parent** must have been **linked**. This means that the **NnRr** parent produced mostly **NR** and **nr gametes**. Some **Nr** and **nR gametes** were still made due to **crossing over**, but there were **fewer Nnrr** and **nnRr** offspring overall. As a result, a higher proportion of the offspring have their parents' phenotypes.

An Epistatic Gene Masks the Expression of Another Gene

1) **Many different genes** can control the **same** characteristic — they **interact** to form the phenotype.
2) This can be because the **allele** of one gene **masks** (blocks) **the expression** of the alleles of other genes — this is called **epistasis**.

Example 1 In humans a **widow's peak** (see picture) is controlled by one gene and **baldness** by others. If you have the **alleles** that code for baldness, it **doesn't matter** whether you have the allele for a widow's peak or not, as you have **no hair**. The baldness genes are **epistatic** to the widow's peak gene, as the baldness genes **mask** the expression of the widow's peak gene.

Example 2 **Flower pigment** in a plant is controlled by two genes. **Gene 1** codes for a **yellow pigment** (Y is the dominant yellow allele) and **gene 2** codes for an enzyme that **turns** the yellow pigment **orange** (R is the dominant orange allele). If you **don't have** the **Y** allele it **won't matter** if you have the R allele or not as the flower **will be colourless**. Gene 1 is **epistatic** to gene 2 as it can **mask** the expression of gene 2.

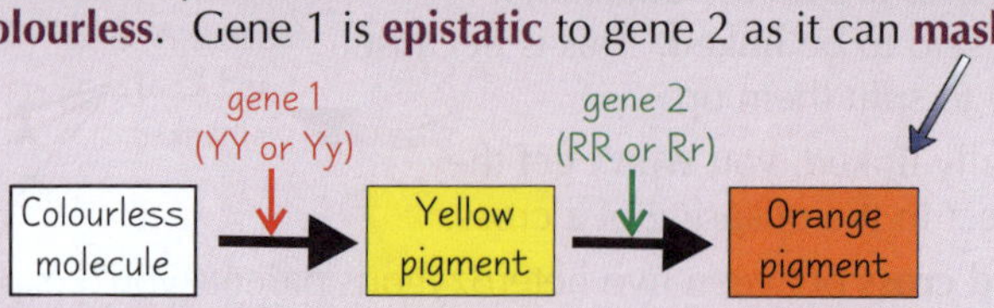

3) **Crosses** involving epistatic genes **don't result** in the **expected phenotypic ratios** given above, e.g. if you cross **two heterozygous orange** flowered plants (YyRr) from the above example you wouldn't get the expected **9 : 3 : 3 : 1** phenotypic ratio for a **normal dihybrid cross**.

Linkage and Epistasis

You can **Predict** the **Phenotypic Ratios** for Some **Epistatic Genes**

Just as you can **predict** the phenotypic ratios for a **normal dihybrid cross** (see page 60), you can predict the phenotypic ratios for dihybrid crosses involving some **epistatic genes** too:

A dihybrid cross involving a recessive epistatic allele — 9 : 3 : 4

Having **two copies** of the **recessive** epistatic allele **masks** (**blocks**) the expression of the **other gene**. If you cross a **homozygous recessive** parent with a **homozygous dominant** parent you will get a **9 : 3 : 4** phenotypic ratio of **dominant both : dominant epistatic recessive other : recessive epistatic** in the F_2 **generation**.

E.g. the **flower example** from the **previous page** is an example of a **recessive epistatic allele**. If a plant is **homozygous recessive** for the **epistatic gene** (**yy**) then it will be **colourless**, **masking** the expression of the orange gene. So if you cross homozygous parents, you should get a **9 : 3 : 4** ratio of **orange : yellow : white** in the F_2 **generation**. You can check the **phenotypic ratio** is right **using a genetic diagram**:

F_1 cross

YYRR × yyrr = all YyRr

F_2 cross: YyRr × YyRr

	YR	Yr	yR	yr
YR	YYRR	YYRr	YyRR	YyRr
Yr	YYRr	YYrr	YyRr	Yyrr
yR	YyRR	YyRr	yyRR	yyRr
yr	YyRr	Yyrr	yyRr	yyrr

Orange = YYRR, YYRr, YyRR, YyRr = 9

Yellow = Yyrr, YYrr = 3

White = yyRR, yyRr, yyrr = 4

Phenotypic ratio: 9 : 3 : 4

A dihybrid cross involving a dominant epistatic allele — 12 : 3 : 1

Having **at least one** copy of the **dominant epistatic** allele **masks** (**blocks**) the expression of the other gene. Crossing a **homozygous recessive** parent with a **homozygous dominant** parent will produce a **12 : 3 : 1** phenotypic ratio of **dominant epistatic : recessive epistatic dominant other : recessive both** in the F_2 generation.

E.g. **squash colour** is controlled by two genes — the **colour epistatic gene** (**W**/**w**) and the **yellow gene** (**Y**/**y**). The **no-colour**, **white** allele (**W**) is **dominant** over the **coloured** allele (**w**), so **WW** or **Ww** will be **white** and **ww** will be **coloured**. The yellow gene has the **dominant yellow** allele (**Y**) and the **recessive green** allele (**y**). So if the plant has **at least one W**, then the squash **will be white**, **masking** the expression of the yellow gene. So if you cross **wwyy** with **WWYY**, you'll get a **12 : 3 : 1** ratio of **white : yellow : green** in the F_2 generation. Here's a **genetic diagram** to prove it:

F_1 cross

WWYY × wwyy = all WwYy

F_2 cross: WwYy × WwYy

	WY	Wy	wY	wy
WY	WWYY	WWYy	WwYY	WwYy
Wy	WWYy	WWyy	WwYy	Wwyy
wY	WwYY	WwYy	wwYY	wwYy
wy	WwYy	Wwyy	wwYy	wwyy

White = WWYY, WWYy, WWyy, WwYY, WwYy, Wwyy = 12

Yellow = wwYY, wwYy = 3

Green = wwyy = 1

Phenotypic ratio: 12 : 3 : 1

Warm-Up Questions

PRACTICE QUESTIONS

Q1 What is a sex-linked characteristic?

Q2 Two genes are autosomally-linked. What does this mean?

Q3 What is an epistatic gene?

Q4 A dihybrid cross produces the phenotypic ratio 9 : 3 : 4 in the F_2 generation. What does this indicate about the genes involved?

Exam Questions

Homozygous curly hair (hhss) crossed with a homozygous bald (HHSS)

Phenotypes of the F_2 offspring produced		
Bald	Straight hair	Curly hair
36	9	3

Q1 Haemophilia A is a sex-linked genetic disorder caused by a recessive allele carried on the X chromosome (X^h).

a) Draw a genetic diagram for a female carrier and a male with haemophilia A to predict the possible genotypes of their offspring. [3 marks]

b) Explain why haemophilia is more common in males than females. [3 marks]

Q2 Hair type in organism A is controlled by two genes: hair (H bald, h hair) and type (S straight, s curly). The F_2 offspring of a cross are shown in the table above.
Use your knowledge of epistasis to explain these results. [3 marks]

Biology students — 9 : 1 phenotypic ratio normal : geek...

I don't know about you but I think I need a lie-down after these pages. Epistasis is a bit of a tricky topic, but you just need to understand what it is and learn the phenotypic ratios for the two types of epistasis — dominant and recessive.

The Chi-Squared Test

Just when you thought it was safe to turn the page... I stick in some maths. Surprise!

The **Chi-Squared Test** Can Be Used to **Check** the **Results** of **Genetic Crosses**

1) The **chi-squared** (χ^2) **test** is a **statistical test** that's used to see if the **results** of an experiment **support** a **theory**.
2) First, the theory is used to **predict** a **result** — this is called the **expected result**. Then, the experiment is carried out and the **actual result** is recorded — this is called the **observed result**.
3) To see if the results support the theory you have to make a **hypothesis** called the **null hypothesis**.
4) The null hypothesis is always that there's **no significant difference** between the observed and expected results (your experimental result will usually be a bit different from what you expect, but you need to know if the difference is just **due to chance**, or because your **theory is wrong**).
5) The χ^2 **test** is then carried out to compare the 'goodness of fit' of the observed and expected results (i.e. to compare how well the observed results match the expected results). The **outcome** either **supports** or **rejects** the **null hypothesis**.
6) You can use the χ^2 test in **genetics** to test theories about the **inheritance** of **characteristics**. For example:

<u>Theory</u>: **Wing length** in fruit flies is controlled by a **single gene** with **two alleles** (**monohybrid inheritance**). The **dominant** allele (N) gives **normal** wings, and the **recessive** allele (n) gives **vestigial** wings.

<u>Expected results</u>: With monohybrid inheritance, if you cross a **homozygous dominant** parent with a **homozygous recessive** parent, you'd expect a **3 : 1 phenotypic ratio** of **normal : vestigial** wings in the F_2 generation (see p. 58).

<u>Observed results</u>: The **experiment** (of crossing a homozygous dominant parent with a homozygous recessive parent) is **carried out** on fruit flies and the **number of F_2 offspring** with normal and vestigial wings is **counted**.

<u>Null hypothesis</u>: There's **no significant difference** between the observed and expected results. (If the χ^2 test shows the observed and expected results are **not significantly different**, then we are **unable to reject** the null hypothesis — the data **supports** the **theory** that wing length is controlled by **monohybrid inheritance**.)

In this kind of statistical test, you can never prove that the null hypothesis is true — you can only 'fail to reject it'. This just means that the evidence doesn't give you a reason to think the null hypothesis is wrong.

First, You Need a **Chi-Squared Value...**

Chi-squared χ^2 is calculated using this formula:
where **O** = **observed** result and **E** = **expected** result.
Σ just means '**the sum of...**'.

$$\chi^2 = \sum \frac{(O-E)^2}{E}$$

Although you won't be expected to calculate a chi-squared value in the written exams, you do need to **understand how the test works**, so that you can **interpret the results**. Here's an example for testing the **wing length** of **fruit flies** as explained above:

Homozygous dominant (NN) flies are crossed with homozygous recessive (nn) flies. **160 offspring** are produced in the F_2 generation.

1. First, the **number of offspring** (out of a total of 160) **expected** for each phenotype is worked out. E for normal wings: 160 (total) ÷ 4 (ratio total) × 3 (predicted ratio for normal wings) = 120. E for vestigial wings: 160 ÷ 4 × 1 = 40.

Phenotype	Ratio	Expected Result (E)	Observed Result (O)
Normal wings	3	120	
Vestigial wings	1	40	

2. Then the **actual number** of offspring **observed** with each phenotype (out of the 160 offspring) is **recorded**, e.g. 111 with normal wings.

Phenotype	Ratio	Expected Result (E)	Observed Result (O)
Normal wings	3	120	111
Vestigial wings	1	40	49

3. The results are used to work out χ^2:

Phenotype	Ratio	Expected Result (E)	Observed Result (O)	O – E	$(O-E)^2$	$\frac{(O-E)^2}{E}$
Normal wings	3	120	111	–9	81	0.675
Vestigial wings	1	40	49	9	81	2.025
					$\Sigma\frac{(O-E)^2}{E}$ =	2.7

The total for this column (2.7) = χ^2

The Chi-Squared Test

...Then Compare it to the Critical Value

1) To find out if there is a **significant difference** between your observed and expected results you need to **compare** the **χ^2 value** to a **critical value**.
2) The critical value is the value of χ^2 that corresponds to a 0.05 (**5%**) level of **probability** that the **difference** between the observed and expected results is **due to chance**.
3) If your χ^2 value is **larger** than or equal to the critical value then there **is a significant difference** between the observed and expected results (something **other than chance** is causing the difference) — and the **null hypothesis** can be **rejected**.
4) If your χ^2 value is **smaller** than the critical value then there **is no significant difference** between the observed and expected results — the null hypothesis **can't be rejected**. E.g. for the example on the previous page the χ^2 value is **2.7**, which is **smaller** than the critical value of **3.84** (see table below) — there's **no significant difference** between the observed and expected results. We've failed to reject the null hypothesis, so the **theory** that wing length in fruit flies is controlled by **monohybrid inheritance** is **supported**.
5) In the exam you might be **given** the **critical value** or asked to **find it** from a **table**:

Using a χ^2 table:

A χ^2 **table** shows a range of **probabilities** that correspond to different **critical values** for different **degrees of freedom** (explained below). Biologists normally use a **probability** level (P value) of **0.05** (5%), so you only need to look in that column.

There's more on P values on p. 117.

- First, the **degrees of freedom** for the experiment are worked out — this is the **number of classes** (number of phenotypes) **minus one**. E.g. 2 – 1 = 1.
- Next, the **critical value** corresponding to a **probability** of **0.05** at **one degree of freedom** is found in the table — here it's **3.84**.
- Then just **compare** your χ^2 value of **2.7** to this critical value, as explained above.

degrees of freedom	no. of classes	Critical values					
1	2	0.46	1.64	2.71	3.84	6.64	10.83
2	3	1.39	3.22	4.61	5.99	9.21	13.82
3	4	2.37	4.64	6.25	7.82	11.34	16.27
4	5	3.36	5.99	7.78	9.49	13.28	18.47
probability that result is due to chance only		0.50 (50%)	0.20 (20%)	0.10 (10%)	0.05 (5%)	0.01 (1%)	0.001 (0.1%)

Warm-Up Questions

PRACTICE QUESTIONS

Q1 What is a χ^2 test used for?

Q2 What can the results of the χ^2 test tell you?

Q3 How do you tell if the difference between your observed and expected results is due to chance?

Exam Question

Q1 A scientist is investigating petal colour in a flower. It's thought to be controlled by two separate genes (dihybrid inheritance), the colour gene — B = blue, b = purple, and the spots gene — W = white, w = yellow. A cross involving a homozygous dominant parent and a homozygous recessive parent should give a 9 : 3 : 3 : 1 ratio in the F_2 generation. The scientist observes the number of offspring showing each of four phenotypes in 240 F_2 offspring. Her results are shown in the table, along with the chi-squared value the scientist calculated for the experiment.

Phenotype	Ratio	Expected Result (E)	Observed Result (O)	$\frac{(O - E^2)}{E}$
Blue with white spots	9	135	131	0.12
Purple with white spots	3	45	52	1.09
Blue with yellow spots	3	45	48	0.20
Purple with yellow spots	1	15	9	2.4
			Chi-squared =	3.81

a) State the null hypothesis for this experiment. [1 mark]

b) The critical value for this experiment is 7.82. Based on the information in the table, is this likely to be a case of dihybrid inheritance or not? Explain your answer. [2 marks]

The expected result of revising these pages — boredom...

...the observed result — boredom. Remember, the null hypothesis (that there's no difference between the observed and expected results) can only be rejected if the value for chi-squared is higher than or equal to the critical value.

The Hardy-Weinberg Principle

Sometimes you need to look at the genetics of a whole population, rather than a cross between just two individuals. And that's where those spiffing fellows Hardy and Weinberg come in...

Members of a Population Share a Gene Pool

1) A **species** is defined as a group of **similar organisms** that can **reproduce** to give **fertile offspring**.
2) A **population** is a group of organisms of the **same species** living in a **particular area** at a **particular time** — so they have the potential to **interbreed**.
3) Species can exist as **one** or **more populations**, e.g. there are populations of the American black bear (*Ursus americanus*) in parts of America and in parts of Canada.
4) The **gene pool** is the complete range of **alleles** present in a **population**.
5) How **often** an **allele occurs** in a population is called the **allele frequency**. It's usually given as a **percentage** of the total population, e.g. 35%, or a **number**, e.g. 0.35.

Yogi wanted everyone to know what population he was in.

The Hardy-Weinberg Principle Predicts Allele Frequencies Won't Change

1) The **Hardy-Weinberg principle** is a mathematical model. It predicts that the **frequencies** of **alleles** in a population **won't change** from **one generation** to the **next**.
2) But this prediction is **only true** under **certain conditions** — it has to be a **large population** where there's **no immigration**, **emigration**, **mutations** or **natural selection**. There also needs to be **random mating** — all possible genotypes can breed with all others.
3) The **Hardy-Weinberg equations** (see below) can be used to **calculate the frequency** of particular **alleles**, **genotypes** and **phenotypes** within populations.
4) The equations can also be used to test whether or not the Hardy-Weinberg principle **applies** to **particular alleles** in **particular populations**, i.e. to test whether **selection** or any **other factors** are **influencing** allele frequencies. — if frequencies **do change** between generations in a large population then there is an influence of some kind.

The Hardy-Weinberg Equations Can be Used to Predict Allele Frequency...

When a gene has two alleles, you can **figure out** the frequency of one of the alleles of the gene if you **know the frequency of the other allele**, using this equation:

$$p + q = 1$$

Where: p = the **frequency** of one allele, usually the **dominant** one
q = the **frequency** of the other allele, usually the **recessive** one

The total frequency of all possible alleles for a characteristic in a certain population is 1.0. So the frequencies of the individual alleles (e.g. the dominant one and the recessive one) must add up to 1.0.

E.g. a species of plant has either **red** or **white** flowers. Allele **R** (red) is **dominant** and allele **r** (white) is **recessive**. If the frequency of **R** is **0.4**, then the frequency of **r** is: 1 – 0.4 = **0.6**.

... Predict Genotype and Phenotype Frequency...

You can **figure out** the frequency of one genotype if you **know the frequencies of the others**, using this equation:

$$p^2 + 2pq + q^2 = 1$$

Where: p^2 = the **frequency** of the **homozygous dominant genotype**
$2pq$ = the **frequency** of the **heterozygous genotype**
q^2 = the **frequency** of the **homozygous recessive genotype**

p^2 is the homozygous dominant genotype frequency if p is the dominant allele.

The total frequency of all possible genotypes for one characteristic in a certain population is 1.0. So the frequencies of the individual genotypes must add up to 1.0.

E.g. if there are **two alleles** for **flower colour** (R and r), there are **three possible genotypes** — **RR**, **Rr** and **rr**. If the frequency of genotype **RR** (**p^2**) is **0.34** and the frequency of genotype **Rr** (**2pq**) is **0.27**, the frequency of genotype **rr** (**q^2**) must be: 1 – 0.34 – 0.27 = **0.39**.

Genotype frequencies can then be used to work out **phenotype frequencies**.

E.g. the frequency of **red flowers** is equal to the genotype frequencies of **RR** and **Rr** added together (0.34 + 0.27 = **0.61**) and the frequency of **white flowers** is equal to the genotype frequency of **rr** (**0.39**).

The Hardy-Weinberg Principle

...Predict the Percentage of a Population that has a Certain Genotype...

EXAMPLE

The **frequency** of **cystic fibrosis** (genotype **ff**) in the UK is currently approximately **1 birth in every 2500**. From this information you can estimate the **percentage** of people in the UK that are cystic fibrosis **carriers** (**Ff**). To do this you need to find the **frequency** of **heterozygous genotype Ff**, i.e. **2pq**, using **both** equations:

$$p + q = 1 \qquad p^2 + 2pq + q^2 = 1$$

First calculate q:
Frequency of cystic fibrosis (homozygous recessive, ff) is 1 in 2500
$ff = q^2 = 1 \div 2500 = 0.0004$
So, $q = \sqrt{0.0004} = 0.02$

Next calculate p:
using $p + q = 1$, $p = 1 - q$
$p = 1 - 0.02 = 0.98$

Then calculate 2pq:
$2pq = 2 \times 0.98 \times 0.02 = 0.039$

The **frequency** of **genotype Ff** is **0.039**, so the **percentage** of the UK population that are **carriers** is **3.9%**.

...and Show if External Factors are Affecting Allele Frequency

EXAMPLE

If the **frequency** of **cystic fibrosis** is measured **50 years later** it might be found to be **1 birth in 3500**. From this information you can estimate the **frequency** of the **recessive allele** (**f**) in the population, i.e. **q**. ⇒

To calculate q:
Frequency of cystic fibrosis (homozygous recessive, ff) is 1 in 3500
$ff = q^2 = 1 \div 3500 = 0.00029$
So, $q = \sqrt{0.00029} = 0.017$

The frequency of the recessive allele is now **0.017**, compared to **0.02** currently (see above).
As the frequency of the allele has **changed** between generations the **Hardy-Weinberg principle doesn't apply** so there must have been some **factors** affecting **allele frequency**, e.g. **immigration**, **emigration**, **mutations** or **natural selection**.

Warm-Up Questions

PRACTICE QUESTIONS

Q1 What is a population?

Q2 What is a gene pool?

Q3 What conditions are needed for the Hardy-Weinberg principle to apply?

Q4 Which term usually represents the frequency of the homozygous recessive genotype in the Hardy-Weinberg equations?

Q5 Which term represents the frequency of the heterozygous genotype in the Hardy-Weinberg equations?

Exam Questions

Q1 Cleft chins are controlled by a single gene with two alleles. The allele coding for a cleft chin (T) is dominant over the allele coding for a non-cleft chin (t). In a particular population the frequency of the homozygous dominant genotype for cleft chin is 0.14.

a) What is the frequency of the recessive allele in the population? *[2 marks]*

b) What is the frequency of the homozygous recessive genotype in the population? *[1 mark]*

c) What percentage of the population have a cleft chin? *[1 mark]*

Q2 In Erminette chickens, feather colour is controlled by a single gene with two codominant alleles — F^B (black feathers) and F^W (white feathers). In a population of Erminette chickens, 43% of birds have the F^B allele. Calculate the frequency of the heterozygous genotype. *[2 marks]*

This stuff's surely not that bad — Hardly worth Weining about...

Two equations that you absolutely have to know — so learn 'em. And whilst you're at it make sure that you learn what each of the terms means as well. You'll feel like a right wally if you know that $p^2 + 2pq + q^2 = 1$ but haven't got a clue what p^2, 2pq and q^2 stand for. It's the kind of stuff that falls out of your head really easily so learn it, learn it, learn it.

Variation and Selection

You might remember a lot of this stuff from Topic 4. Well you need to learn it all again now but with a bit of extra detail for Topic 7. Great. At least there's some extra new stuff to get your teeth stuck into...

Variation Can be **Caused** by **Genes**, the **Environment**, or **Both**

1) **Variation** is the **differences** that exist between individuals.
2) Variation **within a species** means that **individuals** in a population can show a wide range of **different phenotypes**.
3) Although individuals of the **same species** have the **same genes**, they have **different alleles** (versions of genes) — this causes **genetic variation** within a species.
4) The main **source** of this genetic variation is **mutation**, e.g. when changes in the DNA base sequence lead to the production of **new alleles** — see page 84. But genetic variation is also introduced during **meiosis** (through the **crossing over** of chromatids and the **independent segregation** of chromosomes) and because of the **random fertilisation** of gametes during sexual reproduction.
5) Variation within a species can also be caused by differences in the **environment**, e.g. climate, food, lifestyle.
6) Most variation within a species is caused by a **combination** of **genetic** and **environmental** factors. But only **genetic variation** results in **evolution**.

Variation within a species is also called intraspecific variation.

Evolution is a **Change** in **Allele Frequencies Over Time**

The **frequency** of an **allele** in a population **changes** over time — this is **evolution**. **Natural selection** is **one method** by which evolution occurs. Here's a reminder of how it works:

Evolution also occurs by genetic drift. See page 71.

1) **Individuals** of the same species **vary** because they have **different alleles**.
2) **Predation**, **disease** and **competition** (**selection pressures**) create a **struggle for survival**.
3) Because individuals vary, some are **better adapted** to the selection pressures than others.
4) This means there are **differential levels** of **survival** and **reproductive success** in a population. Individuals with a **phenotype** that **increases** their **chance of survival** are **more likely** to **survive**, **reproduce** and **pass on** their genes (including the **beneficial alleles** that determine their phenotype), than individuals with a different phenotype.
5) This means that a **greater proportion** of the next generation **inherit** the **beneficial alleles**.
6) They, in turn, are **more likely** to **survive**, **reproduce** and **pass on** their genes.
7) So the **frequency** of the **beneficial alleles** in the gene pool **increases** from generation to generation.

A selection pressure is anything that affects an organism's chance of survival and reproduction.

Different Types of **Natural Selection** Lead to **Different Frequency Patterns**

Stabilising selection and **directional selection** are **types** of **natural selection** that affect **allele frequency** in different ways. You'll have covered these in Topic 4, but now there's an extra one to learn about — **disruptive selection**.

STABILISING SELECTION is where individuals with alleles for characteristics towards the **middle** of the range are more likely to **survive** and **reproduce**. It occurs when the environment **isn't changing**, and it **reduces the range** of possible **phenotypes**.

Example: In any **mammal population** there's a **range** of **fur length**. In a **stable climate**, having fur at the **extremes** of this range **reduces** the **chances** of **surviving** as it's harder to maintain the **right body temperature**. Animals with alleles for **average fur length** are the **most** likely to **survive**, **reproduce** and **pass on** their alleles. So these alleles **increase** in **frequency**. The **proportion** of the **population** with **average fur length increases** and the **range** of fur lengths **decreases**.

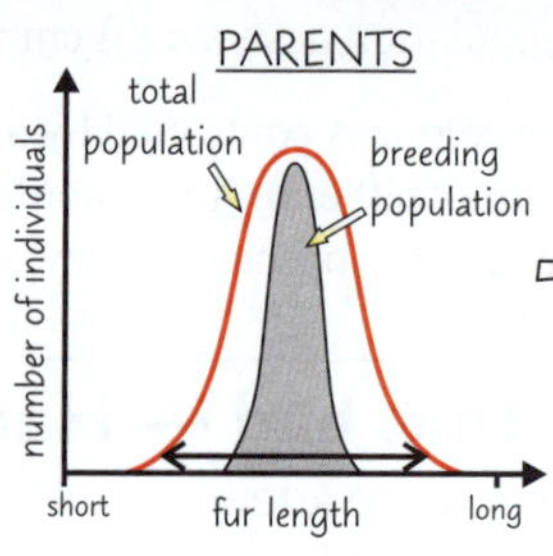

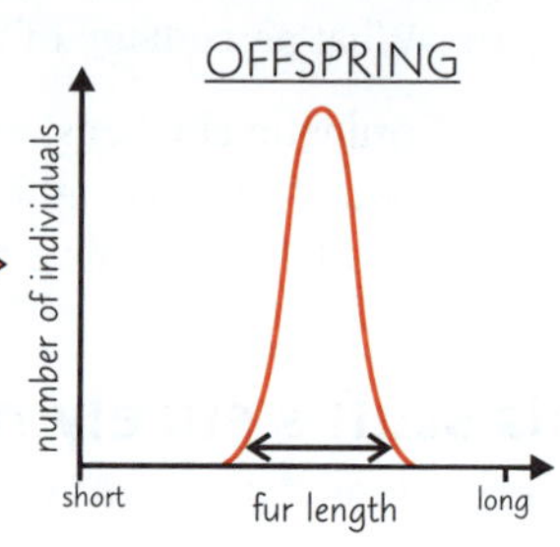

Variation and Selection

DIRECTIONAL SELECTION is where individuals with alleles for a single **extreme phenotype** are more likely to **survive** and **reproduce**. This could be in response to an **environmental change**.

Example: **Cheetahs** are the **fastest** animals on land. It's likely that this characteristic was developed through **directional selection**, as individuals that have **alleles** for increased **speed** are **more likely** to **catch prey** than slower individuals. So they're **more likely** to **survive**, **reproduce** and **pass on** their alleles. Over time the **frequency** of alleles for **high speed increases** and the population becomes **faster**.

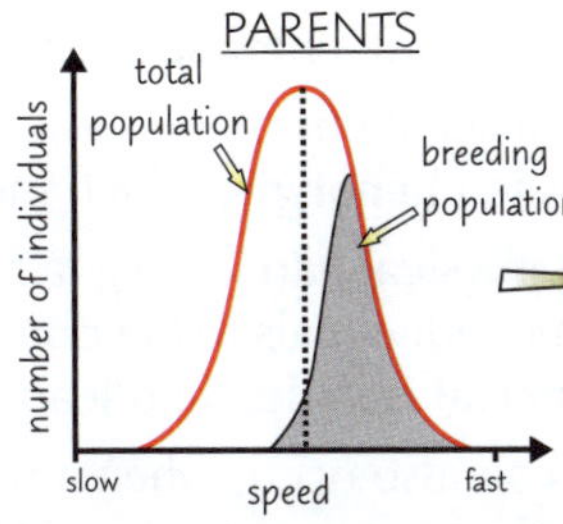

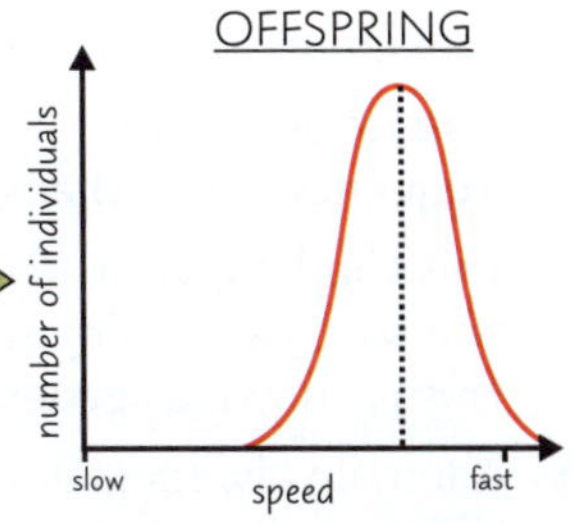

DISRUPTIVE SELECTION is where individuals with alleles for **extreme phenotypes** at **either end of the range** are more likely to **survive** and **reproduce**. It's the **opposite** of **stabilising selection** because characteristics towards the **middle** of the range are **lost**. It occurs when the environment favours **more than one phenotype**.

Example: In **bird populations** there's a **range** of **beak sizes**. Birds with **large beaks** are specialised to eat **large seeds** and birds with **small beaks** are specialised to eat **small seeds**. In an environment where the majority of seeds are large or small and very few (if any) are medium-sized, birds with **medium-sized beaks** may have a **reduced chance of survival**. This is because they are **unable to eat either** large or small seeds effectively. Birds with **large or small** beaks are most likely to **survive**, **reproduce** and **pass on** their alleles. So the alleles for a **large beak** and a **small beak increase in frequency**, but the alleles for a **medium-sized beak decrease in frequency**. Over time the proportion of the population that have **either small or large beaks increases**.

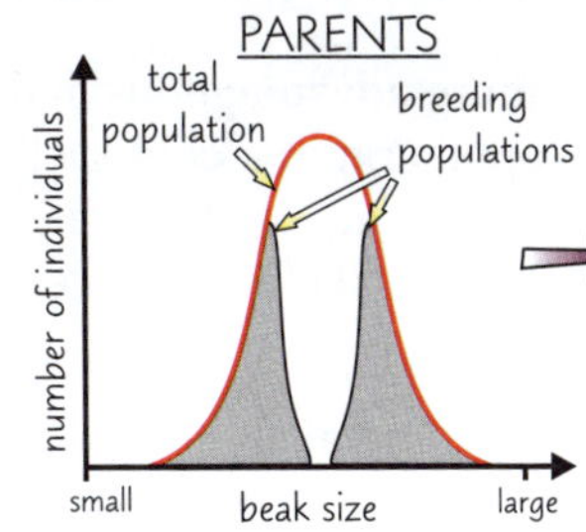

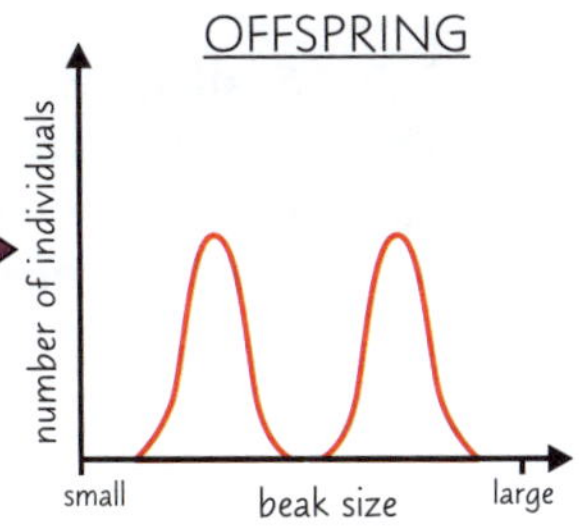

The mating shown here is assortative (non-random) — the birds with small beaks are more likely to mate with other birds with small beaks than they are with large-beaked birds (and vice versa). That's why you end up with two breeding populations.

Warm-Up Questions

PRACTICE QUESTIONS

Q1 Give two possible sources of genetic variation.

Q2 In terms of alleles, what is evolution?

Q3 What is directional selection?

Q4 What is disruptive selection?

Exam Question

Average Temp / °C	Frequency of h allele
22	0.11
21	0.13
19	0.19
18	0.20
16	0.23

Q1 The table on the right shows the results of an investigation into hair length of golden hamsters in a climate where the temperature is decreasing. Hair length is controlled by a single gene with two alleles. H represents the allele for short hair, which is dominant over the allele for long hair, represented by h.

a) Describe the relationship between the frequency of the recessive long hair allele and temperature. Suggest an explanation for this relationship. [4 marks]

b) What type of selection is responsible for this change in allele frequency? [1 mark]

Directional selection — when all the nutty ones are left in a box of chocs...

Ah. more stuff about alleles... it's actually a pretty nice word isn't it? Allele... just rolls off the tongue... Anyway, back to evolution and all that. A key thing to take on board here is that evolution is all about a change in allele (ooh, there it is again) frequency in a population — and natural selection is one way that this can happen.

Speciation and Genetic Drift

Ever wondered how there are so many different species on planet Earth? Well read on and learn, my friend...

Speciation is the **Development** of a **New Species**

1) **Speciation** is the development of a **new species** from an existing species.
2) Speciation occurs when populations of the same species become **reproductively isolated** — **changes in allele frequency** cause **changes in phenotype**, which mean they can **no longer interbreed** to produce **fertile offspring**.
3) This can happen when a **physical barrier**, e.g. a flood or an earthquake, **divides** a population of a species, causing some individuals to become **separated** from the main population. This is known as **geographical isolation** and leads to **allopatric speciation**.
4) Alternatively, speciation can also occur when a population becomes reproductively isolated **without** any **physical separation**. This is known as **sympatric speciation**.

Allopatric Speciation Requires **Geographical Isolation**

1) Populations that are geographically separated will experience slightly **different conditions**. For example, there might be a **different climate** on each side of the physical barrier.
2) This means the populations will experience **different selection pressures** and so **different changes** in allele frequencies could occur:

- Different **alleles** will be **more advantageous** in the different populations. For example, if geographical separation places one population in a **colder climate** than before, **longer fur length** will be **beneficial**. **Directional selection** will then act on the **alleles** for fur length in this population, increasing the frequency of the allele for **longer fur length**.
- Allele frequencies will also change as **mutations** (see p. 84) occur **independently** in each population.
- **Genetic drift** may also affect the allele frequencies in one or both populations (see next page).

3) The changes in allele frequency will lead to **differences** accumulating in the **gene pools** of the separated populations, causing changes in **phenotype frequencies**.
4) Eventually, individuals from the different populations will have changed so much that they won't be able to breed with one another to produce **fertile** offspring — they'll have become **reproductively isolated**.
5) The two groups will have become **separate species**, as shown in the diagram below.

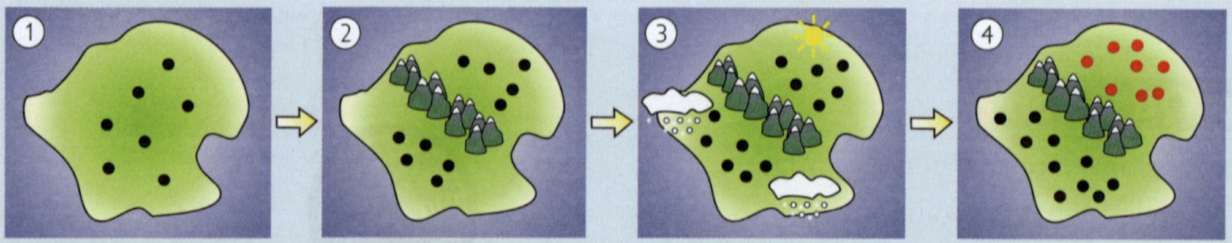

1: Population of individuals. ● = individual organism
2: Physical barriers stop interbreeding between populations.
3: Populations adapt to new environments.
4: Allele and phenotype frequency change, leading to development of new species.

Sympatric Speciation **Doesn't** Require **Geographical Isolation**

A population **doesn't** have to become **geographically isolated** to become **reproductively isolated**. Random mutations could occur **within a population**, **preventing** members of that population breeding with other members of the species.

> There are some more examples of how organisms can become reproductively isolated on the next page.

Example

1) Most eukaryotic organisms are **diploid** — they have **two sets** of **homologous** (matched) **chromosomes** in their cells. Sometimes, **mutations** can occur that **increase** the number of **chromosomes**. This is known as **polyploidy**.
2) Individuals with different numbers of chromosomes **can't reproduce** sexually to give **fertile offspring** — so if a polyploid organism emerges in a diploid population, the polyploid organism will be **reproductively isolated** from the diploid organisms.
3) If the polyploid organism then reproduces **asexually**, a **new species** could develop.
4) Polyploidy can only lead to speciation if it **doesn't prove fatal** to the organism and more polyploid organisms can be produced. It's **more common** in **plants** than animals.

Speciation and Genetic Drift

Reproductive Isolation Occurs in Many Ways

Reproductive isolation occurs because **changes** in alleles and phenotypes in **some individuals prevent** them from **breeding successfully** with individuals **without** these changes. These changes include:

1) **Seasonal** — individuals from the same population develop different **flowering** or **mating** seasons, or become **sexually active** at different times of the year.
2) **Mechanical** — changes in **genitalia** prevent successful mating.
3) **Behavioural** — a group of individuals develop **courtship rituals** that **aren't attractive** to the main population.

Genetic Drift Can Lead to Speciation

1) Different **selection pressures** can change the **allele frequencies** in two geographically isolated species (see previous page). This is evolution by **natural selection**.
2) But evolution can also occur by **genetic drift**. This is when **chance**, rather than **environmental factors**, dictates which individuals **survive**, **breed** and **pass on** their **alleles**:

- Individuals within a population show **variation** in their **genotypes** (e.g. A and B).
- By **chance**, the **allele** for **one genotype** (B) is **passed on** to the offspring **more often** than others.
- So the number of individuals with the allele **increases**.
- **Changes** in allele frequency in two isolated populations could eventually lead to **reproductive isolation** and **speciation**.

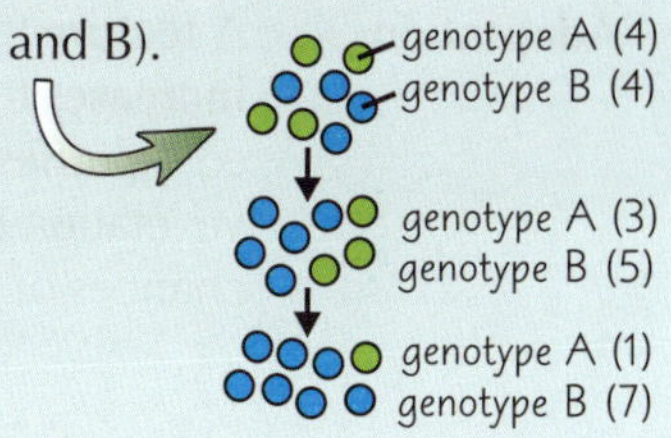

3) Natural selection and genetic drift work **alongside each other** to drive evolution, but one process can drive evolution **more** than the other depending on the **population size**.
4) **Evolution by genetic drift** usually has a **greater effect** in **smaller populations** where **chance** has a **greater influence**. In larger populations, any **chance variations** in allele frequency tend to **even out** across the whole population.

Evolutionary Change Has Resulted in a Great Diversity of Organisms

1) The diversity of life on Earth today is the result of **speciation** and **evolutionary change over millions of years**.
2) To start with there was **one population** of organisms. The population was **divided** and the new populations **evolved** into **separate species**. The new species were then **divided again** and the new populations **evolved** into more separate species.
3) This process has been **repeated** over a long period of time to create millions of new species.

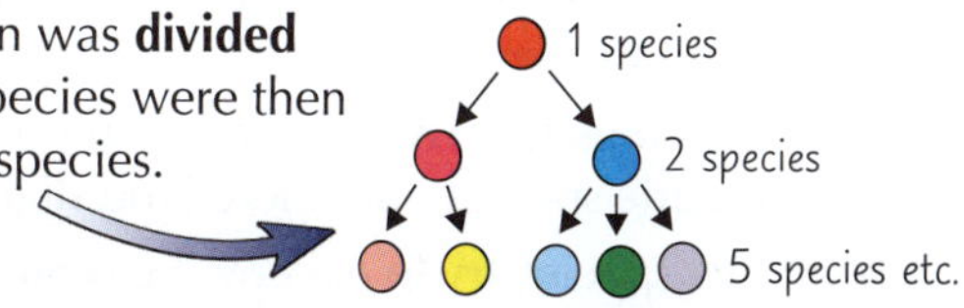

Warm-Up Questions

Q1 What is speciation?

Q2 Describe the process of genetic drift.

PRACTICE QUESTIONS

Exam Question

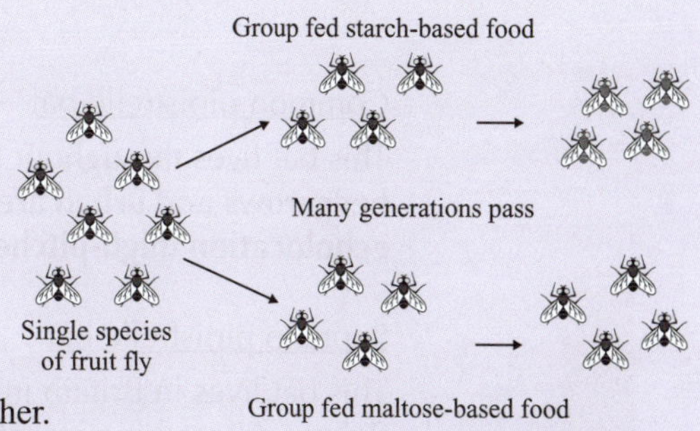

Q1 The diagram shows an experiment conducted with fruit flies. One population was split in two and each population was fed a different food. After many generations the two populations were placed together and it was observed that they were unable to breed together.

a) What evidence shows that speciation occurred? [1 mark]

b) Explain why the experiment resulted in speciation. [3 marks]

If they were ever separated, Al and Patrick would be heartbroken...

So, there are two types of speciation that you need to learn about here. To remember which one allopatric is, I imagine someone on an island shouting 'allo (hello) to their friend Patrick on a separate island. Just thought it might help...

Ecosystems

Ecosystems are amazing — that's why there are all those documentaries about them on TV. You don't have to watch a TV documentary to learn about them though, cos everything you need to know is in this topic.

You Need to **Learn Some Definitions** to get you **Started**

Habitat — The **place** where an organism **lives**, e.g. a rocky shore or a field.

Population — **All** the organisms of **one species** in a **habitat**.

Community — Populations of **different species** in a habitat make up a **community**.

Ecosystem — A **community**, plus all the **non-living** (abiotic) **conditions** in the area in which it lives. Ecosystems can be **small**, e.g. a pond, or **large**, e.g. an entire ocean.

Abiotic conditions — The **non-living** features of the ecosystem, e.g. **temperature** and **availability of water**.

Biotic conditions — The **living** features of the ecosystem, e.g. the presence of **predators** or **food**.

Niche — The **role** of a species within its habitat, e.g. what it eats, where and when it feeds.

Adaptation — A **feature** that members of a species have that **increases** their chance of **survival** and **reproduction**, e.g. **giraffes** have **long necks** to help them reach vegetation that's high up. This increases their chances of survival when food is **scarce**.

Being a member of the undead made it hard for Mumra to know whether he was a living or a non-living feature of the ecosystem.

Every Species Occupies a **Different Niche**

1) The **niche** a species occupies within its habitat includes:
   - Its **biotic** interactions — e.g. the organisms it **eats**, and those it's **eaten by**.
   - Its **abiotic** interactions — e.g. the **oxygen** an organism breathes in, and the **carbon dioxide** it breathes out.
2) Every species has its own **unique niche** — a niche can only be occupied by **one species**.
3) It may **look** like **two species** are filling the **same niche** (e.g. they're both eaten by the same species), but there'll be **slight differences** (e.g. variations in what they eat).
4) If two species **try** to occupy the **same niche**, they will **compete** with each other. One species will be **more successful** than the other, until **only one** of the species is **left**.
5) Here are a couple of examples of niches:

Don't get confused between habitat (where a species lives) and niche (what it does in its habitat).

<u>Common pipistrelle bat</u>

This bat lives throughout Britain on **farmland**, **open woodland**, **hedgerows** and **urban areas**. It feeds by **flying** and catching **insects** using **echolocation** (**high-pitched sounds**) at a **frequency** of around **45 kHz**.

<u>Soprano pipistrelle bat</u>

This bat lives in Britain in **woodland** areas, close to **lakes** or **rivers**. It feeds by **flying** and catching **insects** using **echolocation**, at a **frequency** of **55 kHz**.

It may **look like** both species are filling the **same niche** (e.g. they both eat insects), but there are **slight differences** (e.g. they use **different frequencies** for their echolocation).

Ecosystems

Organisms are Adapted to Biotic and Abiotic Conditions

1) As you know, **adaptations** are features that **increase** an organism's chance of **survival** and **reproduction**.
2) They can be **physiological** (processes **inside** their body), **behavioural** (the way an organism **acts**) or **anatomical** (**structural features** of their body).
3) Organisms with better adaptations are **more likely** to **survive**, **reproduce** and **pass on** the alleles for their adaptations, so the adaptations become **more common** in the population. This is called **natural selection**.
4) Every species is adapted to **use** an **ecosystem** in a way that **no other** species can — it has it's own **unique niche** (see previous page). For example, only giant anteaters can **break into** ant nests and **reach** the ants. They have **claws** to rip open the nest, and a **long**, **sticky tongue** which can move **rapidly** in and out of its mouth to **pick up** the ants.
5) Organisms are **adapted** to both the **abiotic conditions** (e.g. how much **water** is available) and the **biotic conditions** (e.g. what **predators** there are) in their ecosystem.

Here are a few ways that **different organisms** are **adapted** to the **abiotic** or the **biotic** conditions in their ecosystems:

Adaptations to abiotic conditions

- **Otters** have **webbed paws** — this means they can both **walk** on land and **swim** effectively. This increases their chance of survival because they can **live** and **hunt** both on land and in water.
- **Seals** have a **thick layer** of **blubber** (fat) — this helps to keep them **warm** in the **coldest seas**. This increases their chance of survival because they can **live** in places where food is plentiful.
- **Hedgehogs hibernate** — they **lower their rate of metabolism** (all the chemical reactions taking place in their body) over **winter**. This increases their chance of survival because they can **conserve energy** during the **coldest** months.

Adaptations to biotic conditions

- **Sea otters** use **rocks** to **smash open** shellfish and clams. This increases their chance of survival because it gives them **access** to **another source** of food.
- **Male frogs** produce **mating calls** to **attract females** — this makes sure they **attract** a **mate** of the **same species**. This increases their chance of reproduction by making **successful mating** more likely.
- Some **bacteria** produce **antibiotics** — these **kill other species** of bacteria in the **same area**. This increases their chance of survival because there's **less competition** for **resources**.

Warm-Up Questions

Q1 What is the name given to all the organisms of one species in a habitat?
Q2 Define a community.
Q3 Give the term for the non-living features of an ecosystem.
Q4 What happens when two species try to occupy the same niche in an ecosystem?

Exam Question

Q1 Common pipistrelle bats have light, flexible wings, which means they can fly fast and are manoeuvrable. They hunt insects at night using echolocation and live on farmland, in open woodland, hedgerows and urban areas. They make unique mating calls to find mates, hibernate through the winter, and roost in cracks in trees and buildings during the day.

Explain how the common pipistrelle bat is adapted to the biotic conditions in its ecosystem. [3 marks]

Unique quiche niche — say it ten times really fast...

All this population and ecosystem stuff is pretty wordy I'm afraid, but I'll tell you what, you'll be missing it when you get back to the really sciencey stuff later. You just need to learn and relearn all the key words here, then when they ask you to interpret some bat-related babble in the exam, you'll know exactly what they're talking about. Niche work.

Variation in Population Size

Uh-oh, anyone who loves cute little bunnies look away now — these pages are about how the population sizes of organisms fluctuate and the reasons why. One of the reasons, I'm sad to say, is because the little rabbits get eaten.

Population Size Varies Because of Abiotic Factors...

Remember — abiotic factors are the non-living features of the ecosystem.

1) **Population size** is the **total number** of organisms of **one species** in a **habitat**.
2) The **maximum stable population size** of a species that an ecosystem can **support** is called the **carrying capacity**. Carrying capacity varies as a result of both **abiotic** and **biotic factors** (see below).
3) **Abiotic** factors include the amount of **light**, **water** or **space** available, the **temperature** of the surroundings or the **chemical composition** of the surroundings. When abiotic conditions are **ideal** for a species, organisms can **grow fast** and **reproduce successfully**.

   E.g. when the temperature of a mammal's surroundings is the ideal temperature for **metabolic reactions** to take place, they don't have to **use up** as much energy **maintaining** their **body temperature**. This means more energy can be used for **growth** and **reproduction**, so their population size will **increase**.

4) When abiotic conditions **aren't ideal** for a species, organisms **can't** grow as **fast** or reproduce as **successfully**.

   E.g. when the temperature of a mammal's surroundings is significantly **lower** or **higher** than their **optimum** body temperature, they have to **use** a lot of **energy** to maintain the right **body temperature**. This means less energy will be available for **growth** and **reproduction**, so their population size will **decrease**.

...and Because of Biotic Factors

Biotic factors are the living features of the ecosystem.

1 Interspecific Competition — Competition Between Different Species

1) Interspecific competition is when organisms of **different species compete** with each other for the **same resources**, e.g. **red** and **grey** squirrels compete for the same **food sources** and **habitats** in the **UK**.
2) Interspecific competition between two species can mean that the **resources available** to **both** populations are **reduced**, e.g. if they share the **same** source of food, there will be **less** available to both of them. This means both populations will be **limited** by a lower amount of food. They'll have less **energy** for **growth** and **reproduction**, so the population sizes will be **lower** for both species. E.g. in areas where both **red** and **grey** squirrels live, both populations are **smaller** than they would be if there was **only one** species there.
3) If **two** species are competing but one is **better adapted** to its surroundings than the other, the less well adapted species is likely to be **out-competed** — it **won't** be able to **exist** alongside the better adapted species. E.g. since the introduction of the **grey squirrel** to the UK, the native **red squirrel** has **disappeared** from large areas. The grey squirrel has a better chance of **survival** because it's **larger** and can store **more fat** over winter. It can also eat a **wider range** of **food** than the red squirrel.

Never mind what the doctors said, Nutkin knew his weight problem would increase his chance of survival.

2 Intraspecific Competition — Competition Within a Species

Intraspecific competition is when organisms of the **same species compete** with each other for the **same resources**.

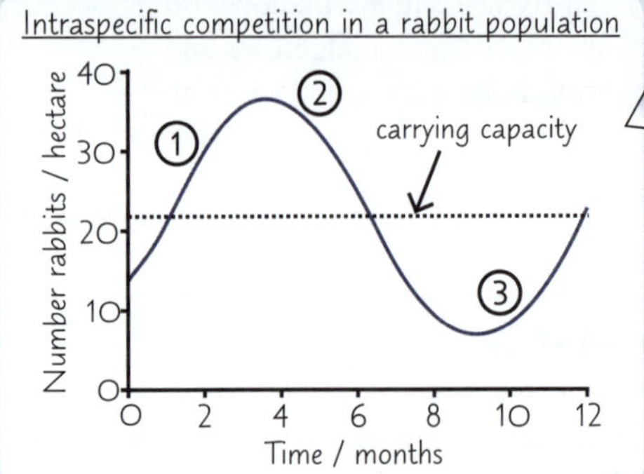

1) The **population** of a species (e.g. rabbits) **increases** when resources are **plentiful**. As the population increases, there'll be **more** organisms competing for the **same amount** of **space** and **food**.
2) Eventually, resources such as food and space become **limiting** — there **isn't enough** for all the organisms. The population then begins to **decline**.
3) A **smaller** population then means that there's **less competition** for space and food, which is **better** for **growth** and **reproduction** — so the population starts to **grow** again.

Variation in Population Size

3 Predation — Predator and Prey Population Sizes are Linked

Predation is where an organism (the predator) kills and eats another organism (the prey), e.g. lions kill and eat (**predate** on) buffalo. The **population sizes** of predators and prey are **interlinked** — as the population of one **changes**, it **causes** the other population to **change**:

1) As the **prey** population **increases**, there's **more food** for predators, so the **predator** population **grows**. E.g. in the graph on the right the **lynx** population **grows** after the **snowshoe hare** population has **increased** because there's **more food** available.
2) As the **predator** population **increases**, **more prey** is **eaten** so the **prey** population then begins to **fall**. E.g. **greater numbers** of lynx eat lots of snowshoe hares, so their population **falls**.
3) This means there's **less food** for the **predators**, so their population **decreases**, and so on. E.g. **reduced** snowshoe hare numbers means there's **less food** for the lynx, so their population **falls**.

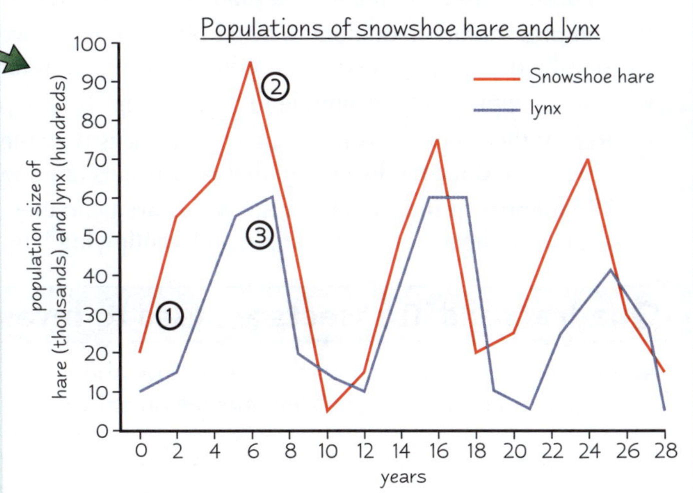

Predator-prey relationships are usually more **complicated** than this though because there are **other factors** involved, like availability of **food** for the **prey**. E.g. it's thought that the population of snowshoe hare initially begins to **decline** because there's **too many** of them for the amount of **food available**. This is then **accelerated** by **predation** from the lynx.

Warm-Up Questions

PRACTICE QUESTIONS

Q1 What is the carrying capacity of an ecosystem?
Q2 Give one example of how an abiotic factor can affect population size.
Q3 What is interspecific competition?
Q4 What will be the effect of interspecific competition on the population size of a species?
Q5 What does it mean when a species is out-competed?
Q6 Define intraspecific competition.

Exam Question

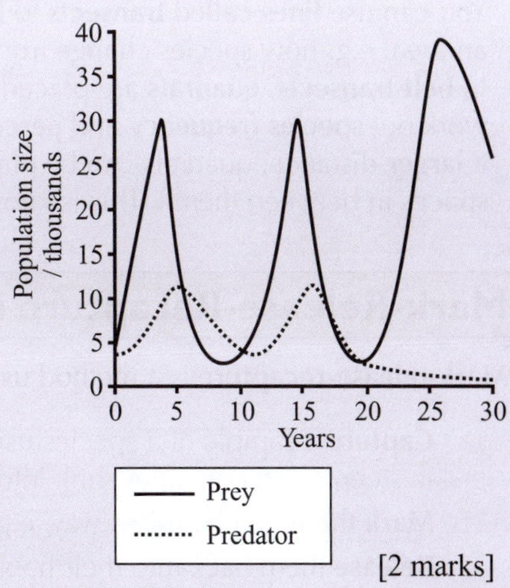

Q1 The graph on the right shows the population size of a prey species and a predator species over a period of 30 years.

a) Calculate the rate at which the prey population increased over the first 4 years. [2 marks]

b) Explain the changes in both the predator and prey populations between years 4 and 10. [3 marks]

c) During what time period was prey population size likely to have been most heavily influenced by intraspecific competition? Give a reason for your answer. [2 marks]

Predator-prey relationships — they don't usually last very long...

You'd think they could have come up with names a little more different than inter- and intraspecific competition. I always remember it as int-er means diff-er-ent species. The factors that affect population size are divided up nicely for you here — just like predators like to nicely divide up their prey into bitesize chunks.

Investigating Populations

PRACTICAL SKILLS

Don't just take my word about all this population stuff — you can go to a field and find out for yourself...

You need to take a **Random Sample** from the **Area You're Investigating**

Most of the time it'd be too **time-consuming** to measure the **number of individuals** in a species (population size) and the **distribution** of that species (i.e. where it's found) in the **entire area** you're investigating. Instead you take **samples**:

1) **Choose** an **area** to **sample** — a **small** area **within** the area being investigated.
2) Samples should be **random** to **avoid bias**, e.g. if you were investigating a field you could pick random sample sites by dividing the field into a **grid** and using a **random number generator** to select **coordinates**.
3) Use an **appropriate technique** to take a sample of the population (see below).
4) **Repeat** the process, taking as many samples as possible.
This will **reduce** the **likelihood** that your results are down to **chance**.
5) The **number of individuals** for the **whole area** can then be **estimated** by taking the **mean** of the data collected in each sample and **multiplying** it by the size of the whole area.

Quadrats and **Transects** are used to **Investigate Non-Motile Organisms**

Non-motile organisms are ones that **don't move** about — like **plants**. Quadrats and transects can also be used to investigate **slow-moving** organisms, which include things like **limpets**.

1) A **quadrat** is a **square** frame, usually divided into a **grid** of 100 **smaller squares** by strings attached across the frame.
2) Quadrats are **placed on the ground** at **different points** within the area you're investigating.
3) The **species frequency** (how often a species is found) or the **number of individuals** of each species is recorded in **each quadrat**.
4) The **percentage cover** of a species can also be measured by counting how much of the quadrat is **covered** by the species — you count a square if it's **more than half-covered**. Percentage cover is a **quick** way to investigate populations and you **don't** have to **count** all the **individual** organisms.

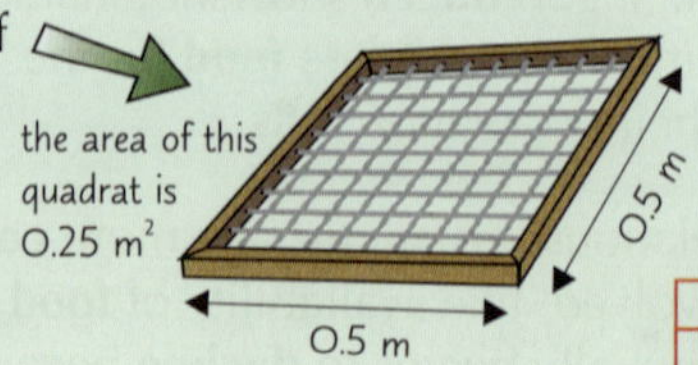

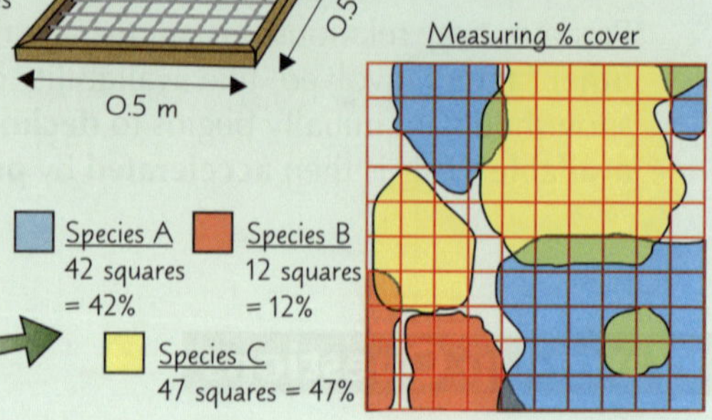

You can use **lines** called **transects** to help find out how plants are **distributed across** an area, e.g. how species change from a hedge towards the middle of a field. In **belt transects**, **quadrats** are placed next to each other **along** the transect to work out **species frequency** and **percentage cover** along the transect. To cover a **larger** distance, quadrats can be placed at **intervals** along the line (i.e. with **spaces** in between them). This is known as an **interrupted** belt transect.

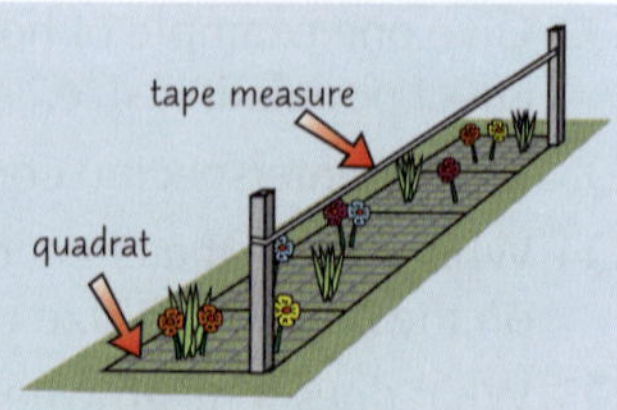

Mark-Release-Recapture is Used to **Investigate More Motile** Species

Mark-release-recapture is a method used to measure the **abundance** of more **motile** species. Here's how it's done:

1) **Capture** a sample of a species using an **appropriate technique**, e.g. you could use pitfall traps (a steep sided container sunk into the ground) to capture ground insects, and **count** them.
2) **Mark** them in a harmless way, e.g. by putting a spot of **paint** on them, or by **removing** a tuft of **fur**.
3) **Release** them back into their habitat.
4) Wait a week, then take a **second sample** from the **same population**.
5) **Count** how many of the second sample **are marked**.
You can then use this **equation** to **estimate** the **total** population size.

$$\text{Total population size} = \frac{\text{Number caught in 1st sample} \times \text{Number caught in 2nd sample}}{\text{Number marked in 2nd sample}}$$

When using this method, you have to make a few **assumptions**:

1) The marked sample has had enough **time** and **opportunity** to **mix** back in with the population.
2) The marking hasn't affected the individuals' **chances of survival** (e.g. by making them more visible to predators), and the marking itself is **still visible** (e.g. it hasn't rubbed off).
3) There are **no changes** in **population size** due to **births**, **deaths** and **migration** during the period of the study.

Investigating Populations

You can Investigate Environmental Factors and Species Distribution

The **distribution** of species often changes within a particular area. E.g. you might find more **shade-loving** plants at the **edge** of a field where they're **sheltered** by a tree, than in the **centre** where they're **exposed** to full sunlight. You need to be able to **investigate** the **effect** of an **environmental factor**, such as shade, on the distribution of a species. Here's an **example** of an investigation you could do to investigate the effect of **soil pH** on **marram grass** in a **coastal ecosystem**:

1) Place a **tape measure** in a straight line from the shore, heading inland. This will be your **transect**.
2) Take a 1 m^2 **quadrat** divided into 100 squares (10 by 10).
3) Starting from the shore, place the quadrat **next to** the tape measure. It doesn't matter where you position the quadrat relative to the tape measure, but you should make sure that you do it the same way each time.
4) **Count** the **squares** containing **marram grass** and record the result in a table as **percentage cover** (as shown below). If you have time, take two repeat quadrat samples next to your initial quadrat and take a mean of your results. Alternatively, you could take a mean of the data from your whole class.
5) At each **sample point**, you should also measure the **pH** and record the results in the table.
6) **Repeat** the observations every 10 m along the transect.

To measure pH:
If you have one, you can use a digital **pH probe** to take pH readings of sand or soil in the field. If you don't have one, you can test the sand/soil back at school. Take a **sample** for testing. When you get back to school, **sieve** it to remove any debris, like twigs and leaves, and place it in a test tube. Add some **barium sulfate**, distilled water and **pH indicator**. **Shake** thoroughly and then leave to it **settle**. Check the **colour** against a **pH chart** and record the result.

Distance from shoreline / m	% cover	pH
0	0	8.5
10	11	8.4
20	27	8.0
30	40	7.6
40	58	7.5
50	55	7.5
60	21	7.1
70	15	7.0
80	8	6.8
90	7	6.6
100	0	6.5

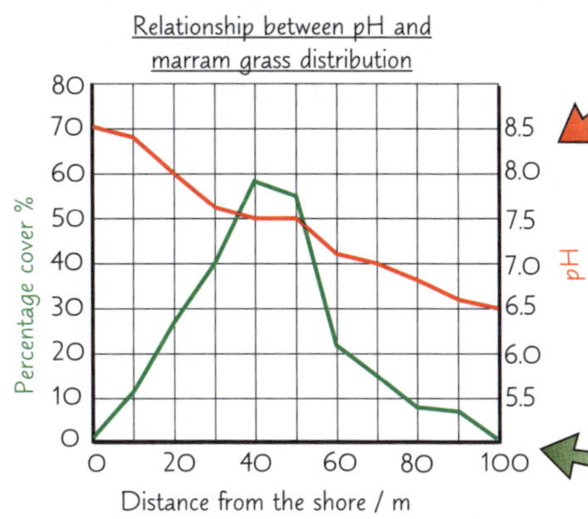

pH decreases as you move **inland**. This is because near the shore the sand/soil contains lots of **shell** fragments which are made of **calcium carbonate**, an **alkaline** compound. Further inland, the **rotting vegetation** adds **organic matter** to the soil, which is more **acidic**.

At first, as **pH decreases** from 8.5 to 7.5, the percentage cover of **marram grass increases**. After pH 7.5, **marram grass** percentage cover **decreases** as **pH** continues to **decrease**. You can't say pH **caused** these trends in marram grass cover though — there could be other factors affecting it, including **soil moisture content**, **salinity**, and **competition** from other species.

Safety issues:
You need to think about what risks you'll be exposed to during fieldwork, so you can plan ways to reduce the chance of them happening. For example, you need to:

- use tide timetables, so you know what the local tide times are when you're working on a beach. Low tide is the best time to work.
- wear suitable clothing and footwear for the weather and terrain, e.g. a sun hat if it's hot and sturdy shoes to stop you slipping.
- wash your hands before eating, especially after handling soil.

Ethical issues:
All fieldwork affects the environment where it's carried out, e.g. lots of people walking around may cause soil erosion and marram grass can be killed by people trampling all over it. Investigations should be planned to have the smallest impact possible, e.g. people should restrict where they walk to the area being studied and try to avoid treading on the plants themselves.

Warm-Up Questions

Q1 Give the formula for calculating population size from the mark-release-recapture method.

An accurate result is one that's close to the true answer (see p. 114).

Exam Question

Q1 A student is investigating the population size of clover plants in a field.
a) Describe how she could estimate the population size of the clover plants using random samples. [4 marks]
b) Explain how incorrect identification of plant species could reduce the accuracy of the results. [1 mark]

What did the quadrat say to the policeman — I've been framed...

If you want to know what it's really like doing these investigations, then read these pages outside in the pouring rain. Doing it while you're tucked up in a nice warm, dry exam hall won't seem so bad after that, take my word for it.

Succession

Repeat after me: successful succession involves several simple successive stages.

Succession is the Process of Ecosystem Change

Ecosystems are **dynamic** — they are constantly **changing**. **Succession** is the process by which an **ecosystem** (see p. 72) **changes** over **time**. The **biotic conditions** (e.g. **plant** and **animal communities**) change as the **abiotic conditions** (e.g. **water** availability) change. There are **two** types of succession:

Remember — biotic = living things, abiotic = non-living.

1) **Primary succession** — this happens on land that's been **newly formed** or **exposed**, e.g. where a **volcano** has erupted to form a **new rock surface**, or where **sea level** has **dropped** exposing a new area of land. There's **no soil** or **organic material** to start with, e.g. just bare rock.
2) **Secondary succession** — this happens on land that's been **cleared** of all the **plants**, but where the **soil remains**, e.g. after a **forest fire** or where a forest has been **cut down by humans**.

Succession Occurs in a Series of Stages

1) **Primary succession** starts when species **colonise** a new land surface. **Seeds** and **spores** are blown in by the **wind** and begin to **grow**. The **first species** to colonise the area are called **pioneer species**.
   - The **abiotic conditions** are **hostile** (**harsh**), e.g. there's no soil to **retain water**. Only pioneer species **grow** because they're **specially adapted** to cope with the harsh conditions, e.g. **marram grass** can grow on sand dunes near the sea because it has **deep roots** to get water and can **tolerate** the salty environment.
   - The pioneer species **change** the **abiotic conditions** — they **die** and **microorganisms decompose** the dead **organic material** (**humus**). This forms a **basic soil**.
   - This makes conditions **less hostile**, e.g. the basic soil helps to **retain water**, which means **new organisms** with **different adaptations** can move in and grow. These then die and are decomposed, adding **more** organic material, making the soil **deeper** and **richer in minerals**. This means **larger plants** like **shrubs** can start to grow in the deeper soil, which retains **even more** water.
   - Some new species may **change** the **environment** so that it becomes **less suitable** for the previous species. E.g. **sand sedge stabilises** the sand through the growth of **rhizomes** (underground stems). This makes the conditions **less suitable** for **marram grass**, which needs constant **reburial** by **sand** in order to grow healthily.
2) **Secondary succession** happens in the **same way**, but because there's already a **soil layer** succession starts at a **later stage** — the pioneer species in secondary succession are **larger plants**, e.g. shrubs.
3) At each stage, **different** plants and animals that are **better adapted** for the improved conditions move in, **out-compete** the plants and animals that are already there, and become the **dominant species** in the ecosystem.
4) As succession goes on, the ecosystem becomes **more complex**. New species move in **alongside** existing species, which means that **biodiversity** (the variety of living organisms) **increases**.
5) The **final stage** is called the **climax community** — the ecosystem is supporting the **largest** and **most complex** community of plants and animals it can. It **won't change** much more — it's in a **steady state**.

This example shows primary succession on bare rock, but succession also happens on sand dunes, salt marshes and even on lakes.

Example of primary succession — bare rock to woodland

1) **Pioneer species colonise** the rocks. E.g. **lichens** grow **on** and **break down** rocks, **releasing minerals**.
2) The lichens **die** and are **decomposed** helping to form a **thin soil**, which thickens as more **organic material** is formed. This means other species such as **mosses** can **grow**.
3) **Larger plants** that need **more water** can move in as the soil **deepens**, e.g. **grasses** and **small flowering plants**. The soil **continues to deepen** as the larger plants die and are decomposed.
4) **Shrubs**, **ferns** and **small trees** begin to grow, **out-competing** the grasses and smaller plants to become the **dominant** species. **Diversity increases**.
5) Finally, the soil is **deep** and **rich** enough in **nutrients** to support **large trees**. These become the dominant species, and the **climax community** is formed.

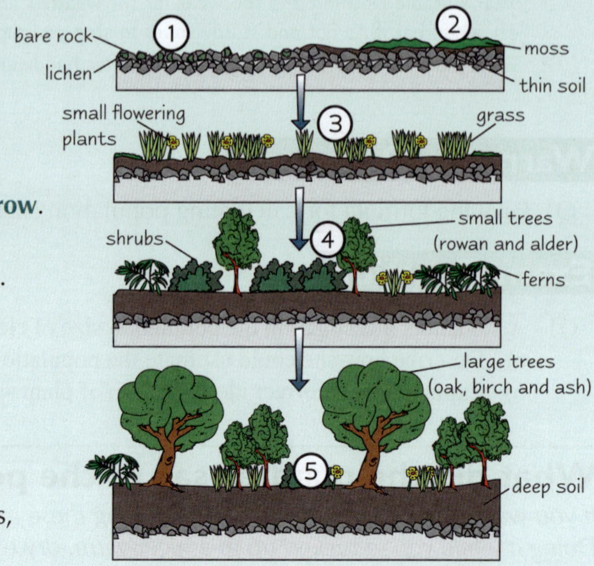

Succession

Different Ecosystems have **Different Climax Communities**

Which species make up the climax community depends on what the **climate** is like in an ecosystem. The climax community for a **particular** climate is called its **climatic climax**. For example:

- In a **temperate climate** there's **plenty** of **available water**, **mild temperatures** and not much **change** between the seasons. The climatic climax will contain **large trees** because they **can grow** in these conditions once **deep soils** have developed.
- In a **polar climate** there's **not much available water**, temperatures are **low** and there are **massive changes** between the seasons. Large trees **won't ever** be able to grow in these conditions, so the climatic climax contains only **herbs** or **shrubs**, but it's still the **climax community**.

Conservation Often **Involves Managing Succession**

Human activities can **prevent succession**, stopping a climax community from **developing**. When succession is stopped **artificially** like this the climax community is called a **plagioclimax**. For example:

A **regularly mown** grassy field **won't develop** shrubs and trees (**woody plants**), even if the climate of the ecosystem could support them. The **growing points** of the woody plants are **cut off** by the lawnmower, so larger plants **can't establish** themselves. The **longer** the interval between mowing, the **further** succession can progress and the more **diversity increases**. But with **more frequent** mowing, succession can't progress and diversity will be **lower** — only the grasses can **survive** being mowed.

A mighty weapon with which to tame the forces of nature.

Conservation (the **protection** and **management** of ecosystems) sometimes involves preventing succession in order to **preserve** an ecosystem in its **current** stage of succession. For example, there are large areas of **moorland** in **Scotland** that provide **habitats** for many species of plants and animals. If the moorland was left to **natural processes**, succession would lead to a **climax community** of **spruce forest**. This would mean the **loss** of the moorland habitat and could lead to the loss of some of the plants and animals that **currently** live there. Preventing succession keeps the moorland ecosystem **intact**. There are a couple of ways to **manage succession** to **conserve** the moorland ecosystem:

1) **Animals** are allowed to **graze** on the land. This is similar to **mowing** — the animals eat the **growing points** of the shrubs and trees, which **stops** them from establishing themselves and helps to keep vegetation **low**.
2) **Managed fires** are lit. After the fires, **secondary succession** will occur on the moorland — the species that grow back **first** (**pioneer species**) are the species that are being **conserved**, e.g. heather. Larger species will take **longer** to grow back and will be **removed again** the next time the moor is burnt.

Warm-Up Questions

Q1 What is the difference between primary and secondary succession?

Q2 What is the name given to species that are the first to colonise an area during succession?

Q3 What is meant by a climax community?

PRACTICE QUESTIONS

Exam Question

Q1 Succession occurs on sand dunes.
You can often see the different stages of succession as you move further inland from the shoreline.

a) Name the type of succession that is taking place when the first grasses start to appear on the dune. Give a reason for your answer. [2 marks]

b) Explain how the growth of grasses can lead to the colonisation of the dune by larger plants like shrubs. [2 marks]

Revision succession — bare brain to a woodland of knowledge...

When answering questions on succession, examiners are pretty keen on you using the right terminology — that means saying "pioneer species" instead of "the first plants to grow there". If you can manage that, then you'll be just fine.

Conservation

Who'd have thought conservation could be such a tricky business — cos I'm feeling nice, I'll try and explain why...

There Can be **Conflict** Between **Human Needs** and **Conservation**

1) Conservation is the **protection** and **management** of **species** and **habitats** (**ecosystems**) in a **sustainable** way. Sustainable means that enough resources are taken to meet the **needs** of people **today**, without **reducing the ability** of people in the **future** to meet their own needs.
2) **Not everyone** agrees with every conservation measure though — there's often **conflict** between **human needs** and **conservation**. **Careful management** is needed to find a **balance** between the two and maintain the **sustainability** of **natural resources**. Here's an example:

- The **Maasai Mara** is a national reserve in **Kenya**. It's a large area of **grassland** (**savannah**) with lots of wildlife.
- The Maasai people traditionally **earn a living** by raising **livestock**, such as **cattle**. This can bring them into **conflict** with conservationists — e.g. **overgrazing** by **livestock** can destroy grassland for **wildlife**.
- Conservation trusts are working with the Maasai to help them **make money** from their land through **conservation** and **ecotourism projects** rather than farming, and to **farm** in a **sustainable way**. So the **economic needs** of the Maasai are met, while still allowing the area and its wildlife to be conserved.

3) There are many **different methods** of conservation. Some focus on conserving a particular **species**, whilst others protect the **habitat** for all the species that live there. Here are some **examples** of conservation techniques:

- Plants can be conserved using **seedbanks**, which are **stores** of lots of **seeds** from lots of **different plant species**. If the plants become **extinct** in the wild, the stored seeds can be used to **grow new plants**.
- **Fishing quotas** are **limits** to the **amount** of certain fish species that fishermen are **allowed** to **catch**. Fishing quotas help to **conserve** fish species by **reducing** the numbers that are **caught** and **killed**.
- **Protected areas** such as **national parks** and **nature reserves** protect habitats (and so protect the **species** in them) by **restricting urban development**, **industrial development** and **farming**.
- **Endangered species** can be **bred** in **captivity** (e.g. a zoo) to **increase** their numbers, then returned to the **wild**.

You May Have to **Evaluate Evidence** and **Data** About **Conservation Issues**

You need to be able to **evaluate** any **evidence** or **data** about **conservation** projects and research that the examiners throw at you — so here's an example I made earlier:

In recent years, **native British bluebells** have become **less common** in woodland areas. It's thought that this is due to the presence of **non-native Spanish bluebells**, which compete with the native species for a **similar niche**. An experiment was carried out to see if **removing** the invasive Spanish species would help to **conserve** the native species. Each year for 15 years the **percentage cover** of native species was estimated in a **50 m by 50 m** area of **woodland** using random sampling and **250, 1 m² quadrats**. After five years, **all** the Spanish bluebells were **removed**. A **similar-sized control woodland** in which the Spanish bluebells remained **untouched** was also studied. The results are shown on the right. You might be asked to:

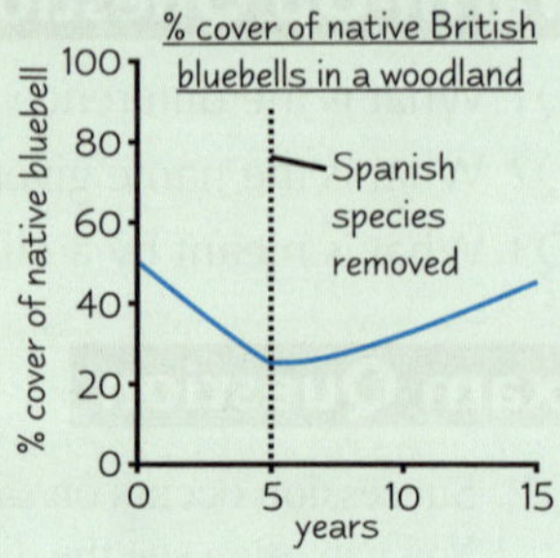

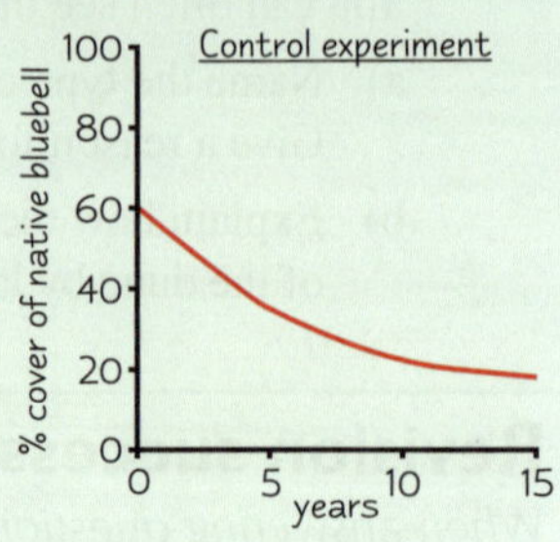

1) **Describe the data:**
   - For the first **five years**, the **percentage cover** of **native bluebells fell** from **50%** to around **25%**. After the Spanish species was **removed**, it **increased** from around **25%** to around **45%** in **ten years**.
   - The **control experiment** shows a fairly **steady drop** in native bluebell percentage cover from **60%** to **20%** over the 15 years.
2) **Draw conclusions:**
   The removal of Spanish bluebells **resulted** in an **increase** in the percentage cover of **native bluebells** over a **ten year period**. This suggests that the **recent decrease** in native British bluebells is due to **competition** with the Spanish bluebells.
3) **Evaluate the method:**
   - The effects of some **other variables** (e.g. **changing weather**) were **removed** by the **control experiment**, where the percentage cover of native bluebells continued to fall throughout the 15-year study. This increases the **validity** of the results.
   - The **study area** and **sample size** were quite **large**, giving **more accurate** data.
   - **Random sampling** removed bias — the data's **more likely** to be an **accurate estimate** of the **whole area**.

Conservation

You Need to be Able to **Consider Conflicting Evidence**

1) The **evidence** from **one study** alone **wouldn't usually be enough** to conclude that there's a **link** between decreasing percentage cover of native bluebells, and the presence of Spanish bluebells.
2) **Similar studies** would be carried out to **investigate** the link. If these studies came to the **same conclusion**, the conclusion would become **increasingly accepted**.
3) Sometimes studies come up with **conflicting evidence** though — evidence that leads to a **different conclusion** than other studies. For example:

Another study was carried out to **investigate** the effect on native bluebells of **removing** Spanish bluebells. It was **similar** to the study above except a **20 m by 20 m** area was sampled using a random sample of **20 quadrats**, and **no control** woodland was used. You might be asked to:

1) **Describe the data:**
In the first five years, the **percentage cover** of **native bluebells fell** from **50%** to around **25%**. After the Spanish species was **removed**, it **kept decreasing** to around **15%** after the **full 15** years.

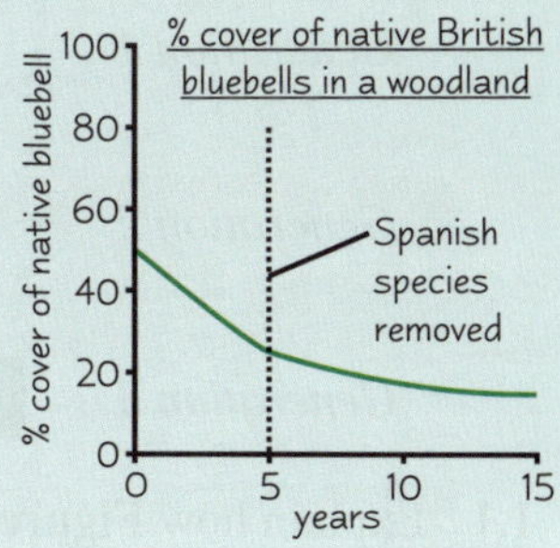

2) **Draw conclusions:**
The **removal** of the Spanish bluebells had **no effect** on the **decreasing** percentage cover of native bluebells — which **conflicts** with the study on the previous page.
3) **Evaluate the method:**
   - There **wasn't** a **control** woodland, so the **continuing decrease** in native bluebell cover after the removal of the Spanish bluebells could be due to **another factor**, e.g. cold weather in years 5-10.
   - The **study area** and **sample size** were quite small, giving a **less accurate** total percentage cover.

Warm-Up Questions

Q1 What is conservation?

Q2 Briefly describe why conflict can occur over conservation issues.

Q3 Suggest one conservation technique that could be used to protect plant species.

Exam Questions

Q1 The graph shows the stock of spawning cod in the North Sea and the rate of mortality caused by fishing from 1963 to 2006.

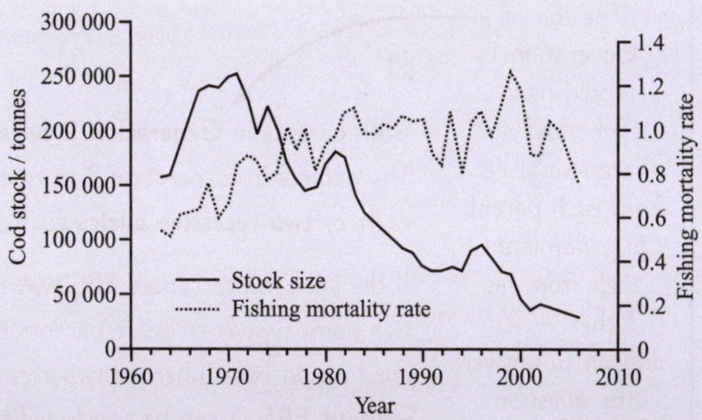

a) Suggest a conclusion that could be drawn from the graph. [2 marks]

b) How might this data be used to make informed decisions about the conservation of cod stocks? [1 mark]

c) Suggest why there might be conflict between conservationists and the North Sea fishing industry. [2 marks]

Q2 Read the following passage and then answer the questions that follow.

Wood, or timber, is an important resource in the UK. It is used in the building industry, to make furniture and as a fuel. Woodlands are also important habitats for many native species. Some deciduous woodland in the UK is managed through a technique called coppicing with standards. When a woodland is managed in this way, just over half of the trees in the woodland are coppiced. This means that the trees are cut down to the stump, and allowed to regrow from shoots which grow from the base of the stump. The rest of the trees are not cut down and are left to grow and mature as normal. These trees are called standards. It's recommended that no more than 40% of the canopy is made up of standard trees.

a) Explain how coppicing (lines 3-5) allows woodland to be managed sustainably. [1 mark]

b) Suggest two benefits of not coppicing all the trees in a woodland. [2 marks]

c) Suggest why it is necessary to restrict how much of the canopy is made up of standards (lines 6-7). [1 mark]

I'm considering conflict after these pages, I tell you...

Ah hah ha, aaaah ha ha ha... oh, I think I need to stop my evil laugh now. Evaluating evidence and data's an important nut to crack — you might have to do it in your exams for conservation, or for another topic altogether.

Extra Exam Practice

It's almost time to crawl back to my favourite niche — the sofa.
Topic 7 is nearly all wrapped up, but there's one more thing you need to do. That's right...
Give these questions mixing parts of Topic 7 a go...

- Have a look at this example of how to answer a tricky exam question.
- Then check how much you've understood from Topic 7 by having a go at the questions on the next page.

Pssst... There are some synoptic questions on pages 123-127 for you to have a go at when you feel confident with all the topics.

1 Epidermolysis bullosa simplex (EBS) is a skin disorder caused by a dominant allele. **Figure 1** shows the incidence of EBS in a family, over three generations. Horizontal lines between two individuals link mating parents and vertical lines link parents and offspring.

Figure 1

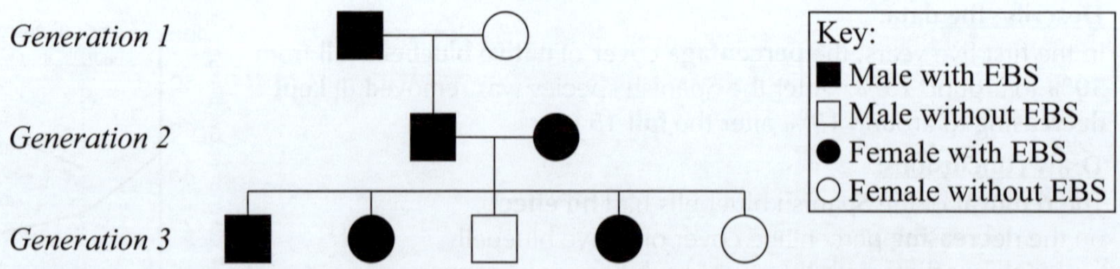

1.1 Explain how **Figure 1** shows that the allele which causes EBS is dominant.

(2 marks)

1.2 It is estimated that 1 in every 17 000 children that are born in the UK, will inherit EBS. Use the Hardy-Weinberg equations to calculate the percentage of people in the UK that are heterozygous for EBS. Show your working.

(3 marks)

The son of Generation 1 could have inherited a recessive allele from each parent, or a dominant allele from his father — it's difficult to answer the question until you look at the offspring of Generation 2.

1.1

Both parents in **Generation 2** have EBS but two of their children don't. The parents in Generation 2 must have, either, at least **one dominant allele** each, or **two recessive alleles** each.

If the allele which causes EBS was recessive then it would be **impossible** for two homozygous recessive parents to have any children without EBS, as each child would have inherited two recessive EBS alleles. As there are **two children without EBS**, it can be concluded that the allele which causes EBS is dominant.

You'd pick up 1 mark for identifying that both parents in Generation 2 must carry a dominant allele and 1 mark for explaining how you know this based on the phenotypes of Generation 3.

1.2

Use the information from the question to work out where to start.

EBS is **coded for by a dominant allele** so children with the disease could have the homozygous dominant genotype or the heterozygous genotype.

So, $p^2 + 2pq = 1 \div 17\,000 = 5.88... \times 10^{-5}$

$\mathbf{p^2 + 2pq + q^2 = 1}$, so $q^2 = 1 - \mathbf{5.88... \times 10^{-5}} = 9.99... \times 10^{-1}$

$q = \sqrt{9.99... \times 10^{-1}} = 9.99... \times 10^{-1}$

$\mathbf{p + q = 1}$, so $p = 1 - 9.99... \times 10^{-1} = 2.94... \times 10^{-5}$

Heterozygotes $= 2pq = 2 \times 2.94... \times 10^{-5} \times 9.99... \times 10^{-1} = 5.88... \times 10^{-5}$

Percentage $= 5.88... \times 10^{-5} \times \mathbf{100} = 0.00588...\% = \mathbf{0.006\%}$

You'd get 3 marks for the correct answer, but if your final answer was wrong you could still pick up 1 mark for correctly identifying $p^2 + 2pq$ as children with EBS, and 1 mark for correctly identifying $2pq$ as heterozygotes.

You should know the Hardy-Weinberg equations and what they mean — see pages 66-67.

Don't round until the end. You can use the 'Ans' button on your calculator to carry the result of one step over to the next.

You've been asked for a percentage so convert from a decimal by multiplying by 100.

Extra Exam Practice

2 **Figure 2** shows the biomass of three different types of plants during succession of a woodland.

Figure 2

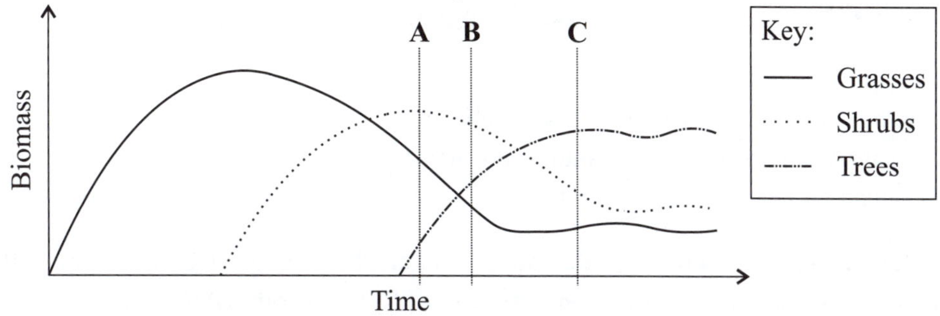

2.1 The grasses shown in **Figure 2** require soil in order to grow.
Explain why this means that **Figure 2** does not show primary succession.

(2 marks)

2.2 Explain how removing trees at point **A** may work as a method of habitat conservation in the woodland.

(2 marks)

2.3 The carrying capacity of several shrub species in the woodland is lower at point **C** than it is at point **B**. Using **Figure 2**, explain why interspecific competition may be responsible for this.

(2 marks)

Wildfires are rapidly spreading fires that can occur naturally in wild areas such as forests. These fires have shaped the evolution of many tree species.

2.4 Trees growing in areas with frequent wildfires often have thicker bark than trees growing in areas with very infrequent wildfires. The thicker bark helps to protect the trees when there is a wildfire. Explain the type of natural selection that may have acted on populations of trees frequently exposed to wildfires.

(3 marks)

A team of scientists wanted to assess the effect of a wildfire on a woodland squirrel population. They used a mark-release-recapture method to sample the squirrels shortly after the wildfire. They also obtained data from similar study that had been done several months before the wildfire. The data they collected is shown in **Figure 3**.

Figure 3

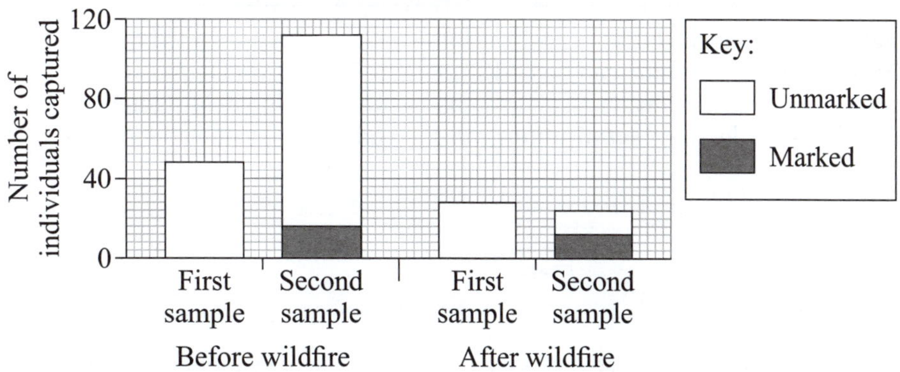

2.5 Before the wildfire, the squirrel population was calculated as 336.
Use **Figure 3** to calculate the percentage change in the population size of the squirrels following the wildfire. Show your working.

(3 marks)

Mutations

Unfortunately, mutations don't usually give you special powers like in superhero movies — in fact, they can be quite harmful. You've already covered mutations in Topic 4, but now you need to know about them in more detail.

Mutations are Changes to the Base Sequence of DNA

1) Any change to the **base (nucleotide) sequence** of DNA is called a **mutation**.
2) Mutations can be caused by **errors** during **DNA replication**.
3) The rate of mutation can be increased by **mutagenic agents** (see next page).
4) The **types** of mutations that can occur include:

- **Substitution** — one or more bases are swapped for another, e.g. ATGCCT becomes ATTCCT.
- **Deletion** — one or more bases are removed, e.g. ATGCCT becomes ATCCT.
- **Addition** — one or more bases are added, e.g. ATGCCT becomes ATGACCT.
- **Duplication** — one or more bases are repeated, e.g. ATGCCT becomes ATGCCCCT.
- **Inversion** — a sequence of bases is reversed, e.g. ATGCCT becomes ACCGTT.
- **Translocation** — a sequence of bases is moved from one location in the genome to another. This could be movement within the same chromosome or movement to a different chromosome.

5) The **order** of **DNA bases** in a gene determines the **sequence of amino acids** in a particular **polypeptide**. If a mutation occurs in a gene, the **sequence** of **amino acids** in the **polypeptide** that it **codes for** could be **changed**.

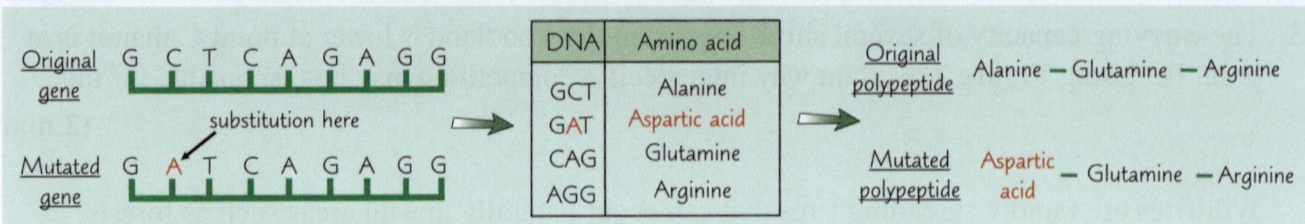

6) Polypeptides make up proteins. A change in the amino acid sequence of a polypeptide may **change** the final **3D shape** of the **protein**, which could mean that it **doesn't work** properly. E.g. a mutation in a polypeptide that makes up an **enzyme** may **change the shape** of the enzyme's **active site**. This may **stop substrates** from being able to **bind to the active site**, leaving the enzyme **unable to catalyse** the reaction.
7) Some mutations can cause **genetic disorders** — inherited disorders caused by **abnormal genes** or **chromosomes**, e.g. cystic fibrosis. Some mutations can **increase** the **likelihood** of developing certain **cancers**, e.g. mutations of the gene **BRCA1** can increase the chances of developing **breast cancer**.
8) If a **gamete** (sex cell) containing a mutation for a genetic disorder or a type of cancer is **fertilised**, the mutation will be present in the new **fetus** formed — these are called **hereditary mutations** because they are passed on to the offspring.

Not all hereditary mutations are harmful — beneficial hereditary mutations drive evolution (see page 68).

Not All Mutations Affect the Order of Amino Acids...

The **degenerate nature** of the genetic code means that some amino acids are coded for by **more than one DNA triplet** (e.g. tyrosine can be coded for by TAT or TAC in DNA). This means **not all** types of mutation will **always** result in a change to the **amino acid sequence** of the **polypeptide**. For example, some **substitutions** will still **code** for the **same amino acid**:

If a mutation doesn't cause a change in the amino acid order, it's called a 'silent mutation'.

Sometimes, **inversion** mutations **don't** cause a **change** in the amino acid **sequence** either:

DNA	Amino acid
TAT	Tyrosine
TAC	Tyrosine
AGT	Serine
CTT	Leucine
CTG	Leucine
TTG	Leucine
GTT	Valine
GTC	Valine

Mutations

...but Some Types of Mutation Do

1) **Additions**, **duplications** and **deletions** within a gene will almost always **change** the **amino acid sequence** of a polypeptide.
2) That's because these mutations all **change** the **number** of **bases** in the DNA code.
3) This causes a **shift** (called a **frameshift**) in the **base triplets** that **follow**, so that the **triplet code** is read in a **different way**.
4) Here's how a **deletion** can cause a **frameshift** and change the amino acid order:

The base triplets that follow on from the mutation are said to be 'downstream' of the mutation.

DNA	Amino acid
TAT	Tyrosine
TAC	Tyrosine
AGT	Serine
CTT	Leucine
GTC	Valine

Mutagenic Agents Increase the Rate of Mutation

Mutations occur **spontaneously**, e.g. when DNA is **misread** during **replication**. But some things can **increase** the **rate** of **mutations** — these are called **mutagenic agents**. **Ultraviolet radiation**, **ionising radiation**, some **chemicals** and some **viruses** are examples of mutagenic agents. They can increase the rate of mutations by:

1) **Acting as a base** — chemicals called **base analogs** can **substitute** for a base during DNA replication, **changing** the **base sequence** in the new DNA. E.g. **5-bromouracil** is a base analog that can substitute for **thymine**. It can pair with **guanine** (**instead** of **adenine**), causing a **substitution mutation** in the new DNA.
2) **Altering bases** — some chemicals can **delete** or **alter bases**. E.g. **alkylating agents** can add an alkyl group to **guanine**, which **changes** the **structure** so that it pairs with **thymine** (**instead** of **cytosine**).
3) **Changing** the **structure of DNA** — some types of **radiation** can change the structure of DNA, which causes **problems** during DNA replication. E.g. **UV radiation** can cause adjacent **thymine** bases to **pair up** together.

It may have been the sunniest summer on record, but nobody expected the extra UV radiation to have such disturbing effects on the pumpkin patch.

Warm-Up Questions

PRACTICE QUESTIONS

Q1 What is a substitution mutation?

Q2 What is the difference between a duplication and an addition mutation?

Q3 What is an inversion mutation?

Q4 What are mutagenic agents?

Q5 List three common mutagenic agents.

Exam Question

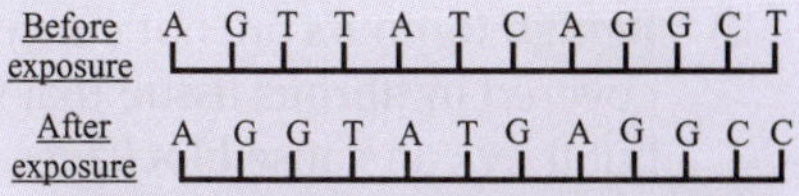

DNA	Amino acids	DNA	Amino acids
AGT	Serine	GAG	Glutamic acid
AGG	Arginine	GCT	Alanine
TAT	Tyrosine	GCC	Alanine
CAG	Glutamine		

Q1 The order of bases in a gene before and after exposure to a mutagenic agent is shown above.

a) Underline any mutation(s) that have occurred. [1 mark]

b) Use the table to explain the changes that the mutations would cause to the sequence of amino acids. [4 marks]

Just hope your brain doesn't have a deletion mutation during the exam...

Right, there's plenty to learn on these pages and some of it's a bit complicated, so you know the drill. Don't read it all through at once — take the sections one by one and get all the facts straight. There could be nothing more fun...

Cancer

Cancer is a disease that affects animals and people of all ages. There are lots of different types of cancer, but they all involve uncontrolled cell growth and all have potentially devastating effects. Here's more on how cancer can occur...

Mutations in **Genes** Can Cause **Uncontrolled Cell Growth**

1) Mutations that occur in individual cells **after** fertilisation (e.g. in adulthood) are called **acquired mutations**.
2) If these mutations occur in the **genes** that **control** the rate of **cell division** (by mitosis), it can cause **uncontrolled cell division**.
3) If a cell divides uncontrollably the result is a **tumour** — a mass of abnormal cells. Tumours that **invade** and **destroy surrounding tissue** are called **cancers** (see below).
4) There are **two types** of **gene** that control cell division — **tumour suppressor genes** and **proto-oncogenes**. Mutations in these genes can cause cancer:

Tumour suppressor genes can be **inactivated** if a **mutation** occurs in the DNA sequence.

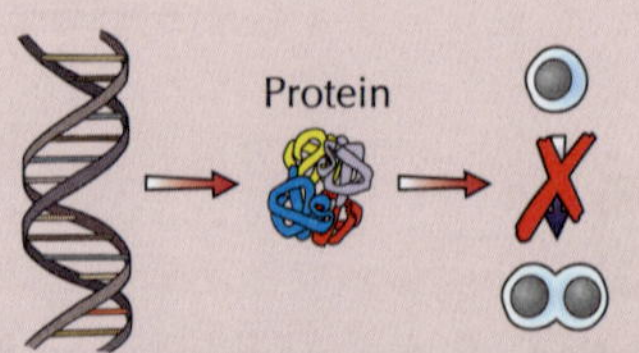

When functioning normally, tumour suppressor genes **slow cell division** by producing proteins that **stop cells dividing** or cause them to **self-destruct** (apoptosis).

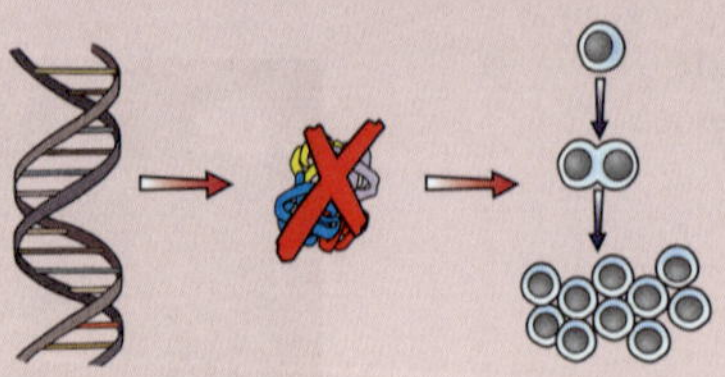

If a **mutation** occurs in a tumour suppressor gene, the protein **isn't produced**. The cells **divide uncontrollably** (the **rate** of division **increases**) resulting in a tumour.

The **effect** of a **proto-oncogene** can be **increased** if a **mutation** occurs in the DNA sequence. A mutated proto-oncogene is called an **oncogene**.

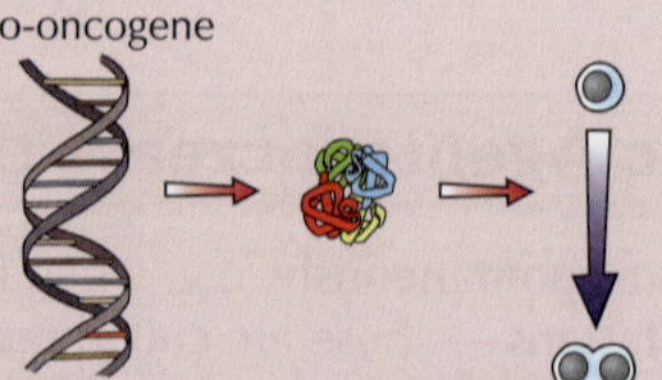

When functioning normally, proto-oncogenes **stimulate cell division** by producing proteins that **make cells divide**.

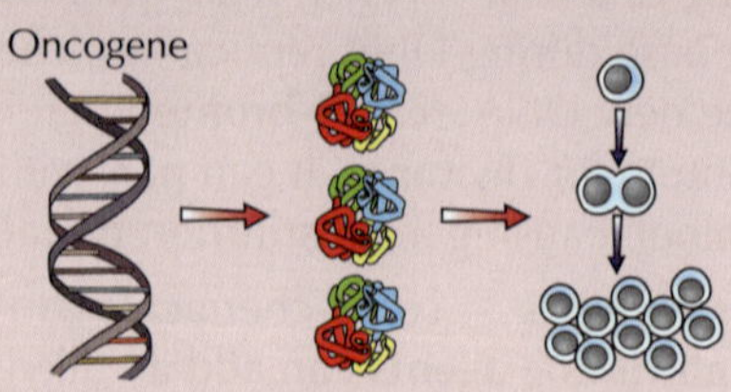

If a **mutation** occurs in a **proto-oncogene**, the gene can become **overactive**. This stimulates the cells to **divide uncontrollably** (the **rate** of division **increases**) resulting in a **tumour**.

Tumours can be **Benign** or **Malignant** (Cancerous)

Tumours can develop for **years** without any obvious symptoms and can be quite **large** by the time they're discovered. **Not** all tumours are **cancerous** — there are **two** different types:

1) **Malignant tumours** are **cancers**. They usually grow **rapidly** and **invade** and **destroy** surrounding tissues. Cells can break off the tumours and **spread** to other parts of the body in the **bloodstream** or **lymphatic system**.
2) **Benign tumours** are not cancerous. They usually grow **slower** than malignant tumours and are often covered in **fibrous tissue** that stops cells invading other tissues. Benign tumours are often **harmless**, but they can cause **blockages** and put **pressure** on organs. Some benign tumours can become **malignant**.

Tumour Cells Look and Function **Differently** to Normal Cells

Tumour cells **can differ** from normal cells in many **different ways**:

1) They have an **irregular shape**.
2) The **nucleus** is **larger** and **darker** than in normal cells. Sometimes the cells have more than one nucleus.
3) They don't produce all the proteins needed to function correctly.
4) They have **different antigens** on their **surface**.
5) They don't respond to **growth regulating processes**.
6) They divide (by mitosis) **more frequently** than normal cells.

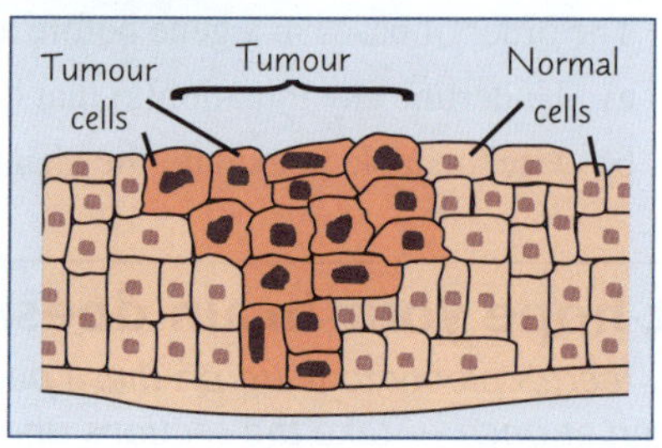

Cancer

Abnormal Methylation of Cancer-Related Genes Can Cause Tumour Growth

1) **Methylation** means **adding** a **methyl** (–CH_3) **group** onto something.
2) **Methylation** of **DNA** is an important method of **regulating gene expression** — it can control **whether** or not a gene is **transcribed** (copied into mRNA) and **translated** (turned into a protein).
3) When methylation is happening **normally**, it plays a **key role** in many processes in the body. It's only when it happens **too much** (**hypermethylation**) or **too little** (**hypomethylation**) that it becomes a **problem**.

For loads more on methylation, see page 97.

4) The **growth** of **tumours** can be **caused** by **abnormal methylation** of certain **cancer-related genes**:

   1) When **tumour suppressor genes** (see previous page) are **hypermethylated**, the genes are **not transcribed** — so the **proteins** they produce to slow cell division **aren't made**. This means that cells are able to **divide uncontrollably** by mitosis and **tumours** can develop.
   2) **Hypomethylation** of **proto-oncogenes** causes them to act as **oncogenes** — **increasing** the **production** of the **proteins** that encourage **cell division**. This stimulates cells to **divide uncontrollably**, which causes the **formation** of **tumours**.

Increased Oestrogen May Contribute to Some Breast Cancers

1) **Increased exposure** to **oestrogen** over an extended period of time is thought to **increase** a woman's **risk** of developing **breast cancer**. (Increased exposure may be the result of starting menstruation earlier than usual or the menopause later than usual. It could also be the result of taking oestrogen-containing drugs, such as HRT.)
2) The **exact** reasons behind this aren't fully understood, but there are a few theories as to how oestrogen can contribute to the **development** of some **breast cancers**:

   1) **Oestrogen** can **stimulate** certain **breast cells** to **divide** and **replicate**. The fact that **more cell divisions** are taking place naturally **increases** the chance of **mutations** occurring, and so **increases** the chance of cells **becoming cancerous**.
   2) This ability to **stimulate division** could also mean that if cells do become **cancerous**, their rapid **replication** could be **further assisted** by **oestrogen**, helping **tumours** to **form quickly**.
   3) Other research suggests that oestrogen is actually able to **introduce mutations directly** into the **DNA** of certain **breast cells**, again **increasing** the chance of these cells becoming **cancerous**.

Warm-Up Questions

Q1 What is a tumour suppressor gene?

Q2 What is the difference between a proto-oncogene and an oncogene?

Q3 What is hypermethylation?

PRACTICE QUESTIONS

Exam Question

Q1 A woman has been diagnosed with cancer. Her doctor has told her that she has a malignant tumour in her left breast.

a) Describe two differences between benign and malignant tumours. [2 marks]

b) Describe how tumours can arise from mutations in DNA. [5 marks]

c) Increased exposure to oestrogen has been linked to some breast cancers. How might oestrogen contribute to causing breast cancer? [4 marks]

Remember, only malignant tumours are cancerous...

You need to understand the difference between benign and malignant. You should never say benign cancers — there's no such thing. Only malignant tumours are cancerous. Make sure that you also know all about the roles that oncogenes and tumour suppressor genes play in causing cancer, as well as the roles of DNA methylation and oestrogen.

Interpreting Data on Cancer

Okay... these pages are a bit daunting. Nevertheless, they're important. Some of the stuff is pretty hard to get your head around, so you'll have to concentrate. After that, take a break and relax. Maybe cut your toenails.

Genetic and **Environmental** Factors Affect the **Risk** of **Cancer**

There's **no single cause** for cancer but scientists have identified lots of different '**risk factors**' — things that **increase** a person's **chance** of getting cancer. Risk factors can be either **genetic** or **environmental**:

1) **Genetic factors** — some cancers are linked with **specific inherited alleles**. If you **inherit** that allele you're **more likely** to get that type of cancer (but it **doesn't mean** you'll **definitely** get that type of cancer).
2) **Environmental factors** — exposure to **radiation**, **lifestyle choices** such as **smoking**, increased **alcohol consumption**, and a **high-fat diet** have all been **linked** to an increased chance of developing some cancers.

It's Difficult to **Interpret** the **Relative Contributions** of **Genes** and **Environment**

1) Data on variation can be very tricky to **interpret** because some characteristics can be affected by **many different genes** (they're polygenic) and **many environmental factors**.
2) It's difficult to know **which factors** (genes or environment) are having the **greatest effect**.
3) This makes it **hard** to **draw conclusions** about the **causes of variation**.

Example: The Effects of Genetic and Environmental Factors on Breast Cancer

Graph to show incidence of breast cancer defined by age and family history

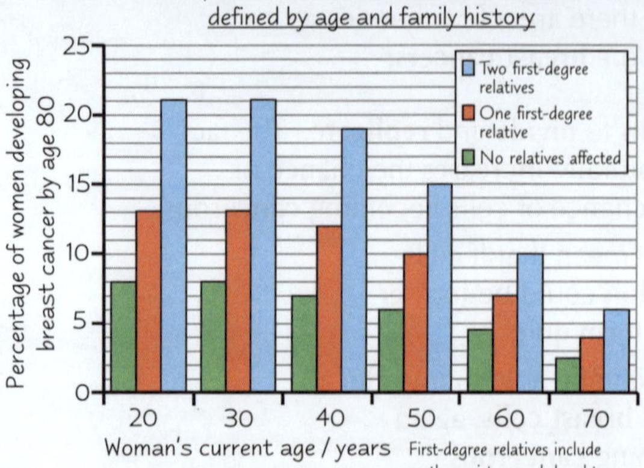

First-degree relatives include mothers, sisters and daughters

This graph shows how the **incidence** of **breast cancer** is affected by both **age** and **family history**. There's a **positive correlation** between **incidence** of breast cancer in women and the number of their **first-degree relatives** who have also had breast cancer. The effect of family history **decreases** with **age**, but the incidence of breast cancer is always **higher** in women with a **close family history** of the disease. A woman is **more likely** to develop breast cancer if **members of her family** have had breast cancer, which suggests a **genetic link**.

Estimated incidence of breast cancer per 100 women according to the number of alcoholic drinks consumed each day

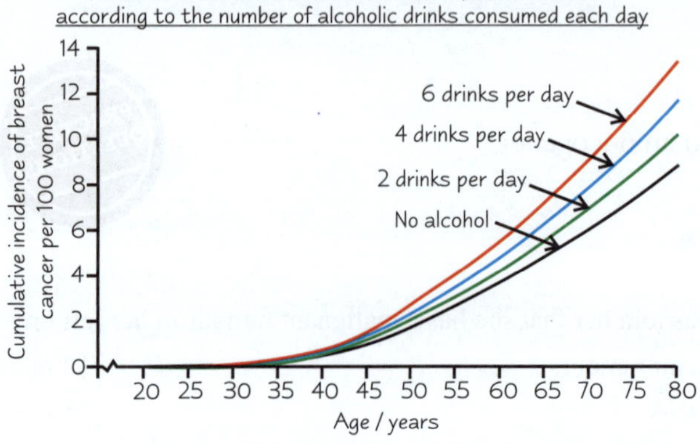

This graph shows that the **incidence** of **breast cancer** is linked to both **age** and **alcohol consumption**. The graph shows that the **incidence** of breast cancer in women **increases** with **age** — i.e. there's a **positive correlation** between incidence of breast cancer and age. There's also a **positive correlation** between the **number** of **alcoholic drinks** consumed **each day** and **incidence** of breast cancer. Alcohol consumption is an **environmental factor.**

1) If you only saw **one** of these graphs you may think **only genetics and age**, or **only alcohol consumption and age**, affect your **risk** of developing **breast cancer**.
2) When you look at **both sets of data** you can see that **all** these things affect the risk.
3) It's **difficult** to tell **which factor** (genes or alcohol) has the **largest effect**.
4) Also, there are **other environmental factors** that are thought to be involved in increasing the risk of developing breast cancer (e.g. **diet**, **exercise**, etc.) that aren't considered here.

There's more on correlations and cause on page 119.

Interpreting Data on Cancer

Knowing the Mutation is Useful for the Prevention and Treatment of Cancer

1) **Cancer** is caused by **mutations** in **proto-oncogenes** and **tumour suppressor genes** (see page 86).
2) **Understanding** the **role** that these genes play in **causing cancer**, and knowing exactly **how** they **work**, can be really **helpful** for coming up with ways to **prevent**, **treat** and **cure** cancer.
3) Here are a few examples:

Prevention

1) If a **specific** cancer-causing **mutation** is **known**, then it is possible to **screen for** (look for) the mutation in a person's DNA (see page 108). E.g. it's possible to screen for the mutated allele of the **BRCA1 tumour suppressor gene**, which greatly increases a woman's risk of developing **breast cancer** in her lifetime.
2) Knowing about this increased risk means that **preventative steps** can be taken to reduce it. E.g. a woman with the BRCA1 mutation may choose to have a **mastectomy** (removal of one or both breasts) to significantly **reduce the risk** of breast cancer **developing**. Women with this mutation may also be **screened** for **signs of breast cancer more often** than the rest of the population, as **early diagnosis increases** the chances of **recovery**.
3) Knowing about specific mutations also means that **more sensitive tests** can be **developed**, which can lead to **earlier** and **more accurate diagnosis**. For example, there's a **mutation** in the **RAS proto-oncogene** in around **half** of all **bowel cancers**. Bowel cancer can be **detected early** by looking for RAS mutations in the DNA of **bowel cells**.

Treatment and Cure

1) The **treatment** for **cancer** can be **different** for different mutations, so knowing how **specific mutations** actually **cause cancer** can be very useful for **developing drugs** to **effectively target** them. For example, **breast cancer** caused by a mutation of the **HER2 proto-oncogene** can be treated with a drug called **Herceptin®**. This drug binds **specifically** to the altered HER2 protein receptor and **suppresses cell division** and **tumour growth**. Breast cancer caused by other mutations is **not** treated with this drug as it doesn't work.
2) Some cancer-causing mutations require **more aggressive treatment** than others, so understanding how the mutation that causes them works can help **produce** the **best treatment plan**. E.g. if a mutation is known to cause an **aggressive** (**fast-growing**) cancer, it may be treated with **higher doses** of **radiotherapy** or by **removing larger areas** of the **tumour** and **surrounding tissue** during surgery.
3) **Gene therapy** (where **faulty alleles** in a person's cells are **replaced** by working versions of those alleles — see page 107) may also be able to treat cancer caused by some mutations. For example, if you know that the cancer is being caused by **inactivated tumour suppressor genes**, it's hoped that gene therapy could be used in the future to provide **working versions** of the genes.

Warm-Up Questions

PRACTICE QUESTIONS

Q1 Give three environmental factors that have been linked to an increased risk of developing cancers.

Q2 How can understanding a specific mutation in a cancer-related gene help treat cancer?

Exam Question

Q1 Possessing a faulty allele of the BRCA1 tumour suppressor gene significantly increases the chance of a woman developing breast cancer in her lifetime. A woman may have her DNA screened for this faulty allele if she has a close family history of breast cancer.

Explain why the ability to screen DNA for the faulty allele may help to prevent a woman with this mutation dying from breast cancer. [4 marks]

Relative contributions — a tenner on your birthday...

In the exam, you might have to evaluate evidence showing correlations between genetic and environmental factors and cancer. Just remember that there are usually several factors at work and that correlation doesn't always mean cause — see page 119 for more. Take a proper look at these examples to help get yourself into the right way of thinking.

Stem Cells

Stem cells — they're the daddy of all cells, the big cheese, the top dog, and the head honcho. And here's why...

Totipotent Stem Cells are Able to Mature into Any Type of Body Cell

1) **Multicellular organisms** are made up from many **different cell types** that are **specialised** for their function, e.g. liver cells, muscle cells, white blood cells.
2) **All** these specialised cell types originally came from **stem cells**.
3) Stem cells are **unspecialised** cells that can develop into **other types** of cell.
4) Stem cells divide to become **new** cells, which then become **specialised**.
5) All multicellular organisms have some form of stem cell.
6) Stem cells are found in the **embryo** (where they become all the **specialised cells** needed to form a **fetus**) and in **some adult tissues** (where they become **specialised** cells that need to be **replaced**, e.g. stem cells in the intestines constantly replace intestinal epithelial cells).
7) Stem cells that can mature (develop) into **any type** of **body cell** in an organism, (including the cells that make up the placenta in mammals) are called **totipotent cells**.
8) **Totipotent** stem cells are **only** present in mammals in the **first few cell divisions** of an **embryo**.
9) After this point the **embryonic stem cells** become **pluripotent**. They can still specialise into **any** cell in the body, but **lose** the **ability** to become the cells that make up the **placenta**.
10) The stem cells present in **adult mammals** are either:

- **Multipotent stem cells** — These are able to differentiate into a **few different types** of cell. For example, both **red** and **white blood cells** can be formed from multipotent stem cells found in **bone marrow**.
- **Unipotent stem cells** — These can only differentiate into **one type** of **cell**. For example, there's a type of unipotent stem cell that can only divide to produce **epidermal skin cells**, which make up the **outer layer** of your **skin**.

Stem Cells Become Specialised Because Different Genes are Expressed

Stem cells become **specialised** because during their development, they only **transcribe** and **translate part** of their **DNA**:

Remember: transcription is when DNA is copied into mRNA. Translation is when proteins are produced using the code in mRNA.

1) **Stem cells** all contain the **same genes** — but during **development**, **not all** of them are **transcribed** and **translated** (expressed).
2) Under the **right conditions**, some **genes** are **expressed** and others are switched off.
3) **mRNA** is only **transcribed** from **specific genes**.
4) The mRNA from these genes is then **translated** into **proteins**.
5) These proteins **modify** the cell — they determine the cell **structure** and **control cell processes** (including the expression of **more genes**, which produces more proteins).
6) **Changes** to the cell produced by these proteins cause the cell to become **specialised**. These changes are **difficult** to **reverse**, so once a cell has specialised it **stays** specialised.

All of the girls expressed different jeans.

Example: Red Blood Cells

1) **Red blood cells** are produced from a type of **stem cell** in the **bone marrow**. They contain lots of **haemoglobin** and have **no nucleus** (to make room for more haemoglobin).
2) The stem cell produces a new cell in which the genes for **haemoglobin production** are **expressed**. Other genes, such as those involved in **removing the nucleus**, are **expressed** too. Many other genes are not expressed (switched off), resulting in a specialised red blood cell.

Stem Cells

Cardiomyocytes Can be Made from **Unipotent Stem Cells**

1) **Cardiomyocytes** are **heart muscle cells** that **make up** a lot of the **tissue** in our hearts. In **mature** mammals, it's thought that they **can't divide** to **replicate** themselves.
2) This meant that for ages, everyone thought that we **weren't able** to **regenerate** our own **heart cells** at all. This is a major problem if the heart becomes **damaged**, e.g. by a **heart attack**, or the cells became **worn out** through age. **Recent research** however, has suggested that our hearts **do** have some **regenerative capability**.
3) Some scientists now think that **old** or **damaged** cardiomyocytes can be **replaced** by **new** cardiomyocytes **derived** from a small supply of **unipotent stem cells** in the heart.
4) Some researchers think that this process could be **constantly occurring**, but haven't yet agreed on **how quickly** it happens.
   - Some believe that it's a **really slow** process and that it's possible that **some** cardiomyocytes are **never replaced** throughout a person's entire **lifetime**.
   - Others think that it's occurring **more quickly**, so that **every** cardiomyocyte in the heart is **replaced several times** in a lifetime.

Stem Cells Can be Used to **Treat Human Disorders**

Since stem cells can divide into a **range** of **specialised cell types**, doctors and scientists think they could be used to **replace** cells **damaged** by illness or injury.

Some **Stem Cell Therapies Already Exist**

1) Some stem cell therapies **already exist** for some diseases affecting the **blood** and **immune system**.
2) **Bone marrow** contains **stem cells** that can become specialised to form **any type** of **blood cell**. **Bone marrow transplants** can be used to replace the **faulty** bone marrow in patients that produce **abnormal blood cells**. The stem cells in the transplanted bone marrow **divide** and **specialise** to produce healthy blood cells.
3) This technique has been used successfully to treat **leukaemia** (a **cancer** of the blood or bone marrow) and **lymphoma** (a cancer of the **lymphatic system**).
4) It has also been used to treat some **genetic disorders**, such as **sickle-cell anaemia** and **severe combined immunodeficiency** (**SCID**):

Example

Severe combined immunodeficiency (**SCID**) is a genetic disorder that affects the immune system. People with SCID have a **poorly functioning immune system** as their **white blood cells** (made in the bone marrow from stem cells) are **defective**. This means they **can't defend** the body against infections by identifying and destroying microorganisms. So SCID sufferers are **extremely susceptible** to **infections**. Treatment with a **bone marrow transplant** replaces the faulty bone marrow with donor bone marrow that contains **stem cells without** the **faulty genes** that cause SCID. These then **differentiate** to produce **functional** white blood cells. These cells can identify and destroy invading pathogens, so the **immune system functions properly**.

Stem Cells Could be Used to **Treat Other Diseases**

Scientists are **researching** the use of stem cells as **treatment** for lots of conditions, including:

- **Spinal cord injuries** — stem cells could be used to replace damaged **nerve tissue**.
- **Heart disease** and **damage caused by heart attacks** — stem cells could be used to replace damaged **heart tissue**.
- **Bladder conditions** — stem cells could be used to grow **whole bladders**, which are then **implanted** in patients to replace diseased ones.
- **Respiratory diseases** — **donated windpipes** can be stripped down to their simple collagen structure and then covered with **tissue** generated by stem cells. This can then be **transplanted** into patients.
- **Organ transplants** — organs could be **grown** from stem cells to provide new organs for people on **donor waiting lists**.

These treatments aren't available yet but some are undergoing clinical trials.

Stem Cells

There are **Huge Benefits** to Using **Stem Cells** in **Medicine**

People who make **decisions** about the **use** of stem cells to treat human disorders have to consider the **potential benefits** of stem cell therapies:

- They could **save** many **lives** — e.g. many people waiting for organ transplants **die** before a **donor organ** becomes available. Stem cells could be used to **grow organs** for those people awaiting transplants.
- They could **improve** the **quality of life** for many people — e.g. stem cells could be used to replace damaged cells in the eyes of people who are **blind**.

Human **Stem Cells** Can Come from **Adult Tissue** or **Embryos**

To **use stem cells** scientists have to get them from somewhere. There are **three** main potential **sources** of human stem cells:

1 Adult Stem Cells

1) These are obtained from the **body tissues** of an **adult**. For example, adult stem cells are found in **bone marrow**.
2) They can be obtained in a relatively **simple operation** — with very **little risk** involved, but quite **a lot** of **discomfort**.
3) Adult stem cells **aren't** as **flexible** as embryonic stem cells — they can only specialise into a **limited** range of cells, not all body cell types (they're **multipotent**).

2 Embryonic Stem Cells

1) These are obtained from **embryos** at an **early stage of development**.
2) Embryos are created in a **laboratory** using ***in vitro* fertilisation** (IVF) — **egg cells** are **fertilised** by sperm **outside the womb**.
3) Once the embryos are approximately **4 to 5 days old**, **stem cells** are **removed** from them and the rest of the embryo is **destroyed**.
4) Embryonic stem cells can divide an **unlimited number** of times and develop into **all types** of body cells (they're **pluripotent**).

3 Induced Pluripotent Stem Cells (iPS Cells)

1) iPS cells are created by scientists in the **lab**. The process involves **'reprogramming' specialised adult body cells** so that they **become pluripotent**.
2) The adult cells are made to **express** a series of **transcription factors** that are normally associated with pluripotent stem cells. The transcription factors cause the adult body cells to **express genes** that are associated with pluripotency.
3) One of the ways that these transcription factors can be **introduced** to the **adult cells** is by **infecting** them with a **specially-modified virus**. The virus has the **genes coding** for the transcription factors within its DNA. When the virus **infects** the **adult cell**, these **genes** are passed **into** the **adult cell's DNA**, meaning that the cell is **able** to **produce** the **transcription factors**.
4) Induced pluripotent stem cells could become really useful in **research** and **medicine** in the future — see next page. At the moment though, **more research** into how **similar** they actually are to **true pluripotent embryonic stem cells** is needed before they can be properly utilised.

Transcription factors are proteins that control whether or not genes are transcribed — see page 94 for more.

Stem Cells

There are Ethical Issues Surrounding Embryonic Stem Cell Use

1) Obtaining stem cells from **embryos** created by IVF raises **ethical issues** because the procedure results in the **destruction** of an embryo that **could** become a **fetus** if placed in a **womb**.
2) Some people believe that at the moment of **fertilisation** an **individual** is formed that has the **right** to **life** — so they believe that it's **wrong** to **destroy** embryos.
3) Some people have **fewer objections** to stem cells being **obtained** from **egg cells** that **haven't** been fertilised by sperm, but have been **artificially activated** to start **dividing**. This is because the cells **couldn't survive** past a few days and **wouldn't** produce a fetus if placed in a womb.
4) Some people think that **scientists** should **only use** adult stem cells because their production **doesn't** destroy an embryo. But adult stem cells **can't** develop into all the specialised cell types that embryonic stem cells can.
5) This is where **induced pluripotent stem cells** could prove really useful. They have the potential to be **as flexible** as **embryonic stem cells**, but, as they're **obtained** from **adult tissue**, there **aren't** the same **ethical issues** surrounding their use. Good news all round.
6) It's also possible that iPS cells could be made from a **patient's own cells**. These iPS cells, which would be **genetically identical** to the patient's cells, could then be used to **grow** some **new tissue** or **an organ** that the patient's body **wouldn't reject** (rejection of transplants occurs quite often and is caused by the patient's immune system recognising the tissue as **foreign** and **attacking it**).
7) The decision makers in **society** have to take into account **everyone's views** when making decisions about **important scientific work** like stem cell research and its use to treat human disorders.

You might be asked to evaluate the use of stem cells in treating human disorders in the exams — so make sure you know all the pros and cons on pages 91 to 93.

Warm-Up Questions

Q1 At what stage of development can totipotent stem cells be found in mammals?

Q2 Describe the difference between pluripotent and multipotent stem cells.

Q3 How do stem cells become specialised?

Q4 Name two conditions that stem cells could potentially be used to treat.

Q5 Describe one difference between embryonic and adult stem cells.

Tina, Joe and Bex knew their cells were specialised — specialised to look good.

PRACTICE QUESTIONS

Exam Question

Q1 Scientists are currently exploring the potential for the use of stem cells in medicine.

a) Explain one way in which stem cell therapy is currently being used. [4 marks]

b) Explain why some people object to the use of embryonic stem cells in treating human disorders. [2 marks]

It may be possible to use induced pluripotent stem cells (iPS cells) instead of embryonic stem cells to treat human disorders.

c) Describe how induced pluripotent stem cells can be produced. [4 marks]

It's OK — you can grow yourself a new brain especially for this revision...

Stem cells are pretty amazing when you think about it — some can differentiate into absolutely any cell type needed to form an organism. I guess that makes them the cellular equivalent to those giant penknives that have a tool for everything. Totipotent stem cells are the most flexible, followed by pluripotent, multipotent, then unipotent stem cells. Some stem cells are already being used in medicine, but their full potential isn't currently being met. You need to be able to evaluate the use of stem cells in medicine, taking all the benefits and drawbacks on these pages into account.

Regulation of Transcription and Translation

These pages cover some of the ways that transcription and translation are regulated. It's really all incredibly clever.

Transcription Factors Control the Transcription of Target Genes

You should remember from Topic 4 that **transcription** is when a **gene** is **copied** from DNA into **messenger RNA** (**mRNA**). The enzyme responsible for synthesising mRNA from DNA is called **RNA polymerase**.

1) **All** the **cells** in an organism carry the **same genes** (DNA) but the **structure** and **function** of different cells **varies**.
2) This is because **not all** the **genes** in a cell are **expressed** (transcribed and used to make a protein).
3) Because **different genes** are expressed, **different proteins** are made and these proteins modify the cell — they determine the **cell structure** and control **cell processes** (including the expression of more genes, which produce more proteins).
4) The **transcription** of genes is **controlled** by protein molecules called **transcription factors**:

1) In eukaryotes, transcription factors **move** from the **cytoplasm** to the **nucleus**.
2) In the nucleus they **bind** to **specific DNA sites** near the start of their **target genes** — the genes they **control** the expression of.
3) They control expression by controlling the **rate** of transcription.
4) Some transcription factors, called **activators**, **stimulate** or **increase** the **rate of transcription** — e.g. they help **RNA polymerase bind** to the start of the target gene and **activate** transcription.
5) Other transcription factors, called **repressors**, **inhibit** or **decrease** the **rate of transcription** — e.g. they **bind** to the start of the target gene, **preventing RNA polymerase** from **binding**, **stopping** transcription.

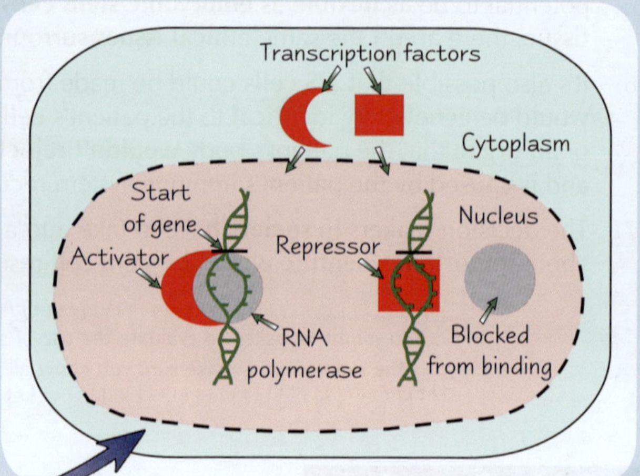

Oestrogen Can Initiate the Transcription of Target Genes

The **expression** of **genes** can also be **affected** by **other molecules** in the cell, e.g. **oestrogen**:

1) Oestrogen is a steroid **hormone** that can affect transcription by **binding** to a **transcription factor** called an **oestrogen receptor**, forming an **oestrogen-oestrogen receptor complex**.
2) The complex moves from the **cytoplasm** into the **nucleus** where it **binds** to **specific DNA sites** near the **start** of the **target gene**.
3) The complex can act as an **activator** of transcription, e.g. **helping** RNA polymerase bind to the start of the target gene.

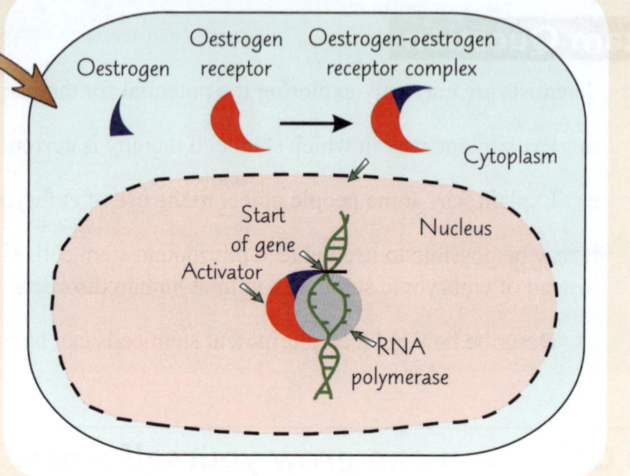

In some cells, the oestrogen-oestrogen receptor complex can act as a repressor of transcription instead of an activator. It depends on the type of cell and the target gene.

Regulation of Transcription and Translation

RNA Interference (RNAi) Can Inhibit the Translation of mRNA

1) In **eukaryotes**, **gene expression** is also affected by **RNA interference** (**RNAi**).
2) RNAi is where small, double-stranded **RNA** molecules **stop mRNA** from target genes being **translated** into **proteins**. A **similar process** to RNAi can also occur in prokaryotes.
3) The molecules involved in RNAi are called **siRNA** (small interfering RNA) and **miRNA** (microRNA).
4) Here's how RNAi works:

RNAi molecules are small lengths of non-coding RNA (they don't code for proteins).

siRNA (and miRNA in plants)

1) Once mRNA has been transcribed, it leaves the nucleus for the **cytoplasm**.
2) In the cytoplasm, double-stranded **siRNA** associates with several **proteins** and unwinds. A single strand then **binds** to the **target mRNA**. The **base sequence** of the **siRNA** is **complementary** to the base sequence in sections of the **target mRNA**.
3) The **proteins** associated with the **siRNA cut** the mRNA into fragments — so it can **no longer** be **translated**. The fragments then move into a **processing body**, which contains 'tools' to **degrade** them.
4) A **similar process** happens with **miRNA** in **plants**.

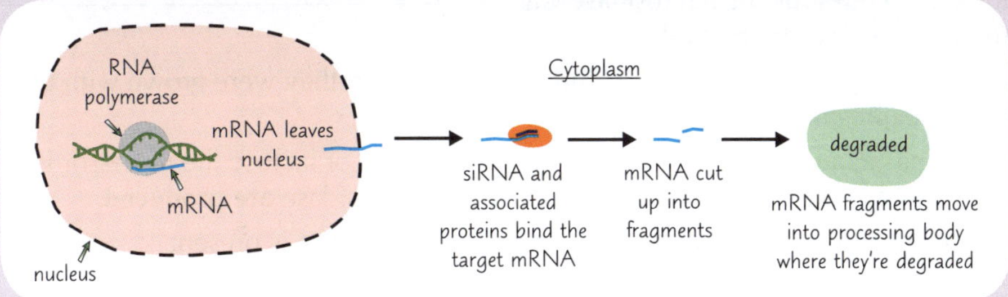

miRNA in mammals

1) In **mammals**, the **miRNA isn't** usually **fully complementary** to the target mRNA. This makes it **less specific** than siRNA and so it may target **more than one** mRNA molecule.
2) Like siRNA, it associates with proteins and **binds** to **target mRNA** in the **cytoplasm**.
3) Instead of the proteins associated with miRNA cutting mRNA into fragments, the miRNA-protein complex physically **blocks** the **translation** of the **target mRNA**.
4) The mRNA is then moved into a **processing body**, where it can either be **stored** or **degraded**. When it's stored, it can be **returned** and **translated** at **another time**.

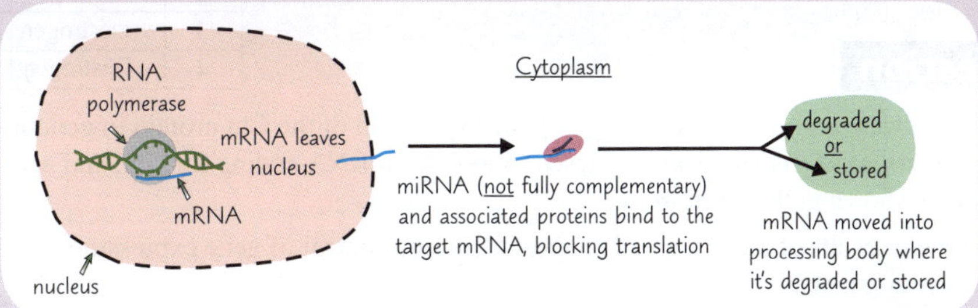

You Need to be Able to Interpret Experimental Data on Gene Expression

You could get a question in the exam where you have to **interpret data** about **gene expression**. It could be on anything you've learnt about on these two pages (e.g. **transcription factors**, **oestrogen** or **RNAi**) or it could be on **epigenetic control** of gene expression (see pages 97-98).

On the next page there's an example of a **gene expression system** in bacteria and an experiment that **investigates** how it works. You **don't** need to **learn** the information, just **understand** what the results of the experiment tell you about how the expression of the gene is **controlled**.

Regulation of Transcription and Translation

The *lac* repressor:

1) ***E. coli*** is a **bacterium** that respires **glucose**, but it can use **lactose** if glucose **isn't available**.
2) If lactose is present, *E. coli* makes an **enzyme** (**β-galactosidase**) to **digest** it. But if there's **no** lactose, it doesn't **waste energy** making an enzyme it **doesn't need**. The enzyme's **gene** is **only expressed** when lactose is **present**.
3) The production of the enzyme is **controlled** by a **transcription factor** — the ***lac* repressor**.
4) When there's **no** lactose, the *lac* repressor **binds** to the **DNA** at the start of the gene, **stopping transcription**.
5) When lactose **is present** it **binds** to the *lac* repressor, **stopping** it binding to the DNA, so the gene **is transcribed**.

Experiment:

1) Different ***E. coli*** mutants were isolated and grown in **different media**, e.g. with lactose or glucose.
2) The mutants have **mutations** (**changes** in their **DNA bases**, see page 84) that mean they **act differently** from normal *E. coli*, e.g. they **produce β-galactosidase** when grown with glucose.
3) To **detect** whether active (working) β-galactosidase was produced, a **chemical** that turns **yellow** in the presence of active β-galactosidase was **added** to the medium.
4) The production of **mRNA** that **codes for β-galactosidase** was also measured. The results are shown in the **table**.

Medium	Mutant	mRNA	Colour
Glucose	Normal	No	No yellow
Lactose	Normal	Yes	Yellow
Glucose	Mutant 1	Yes	Yellow
Lactose	Mutant 1	Yes	Yellow
Glucose	Mutant 2	No	No yellow
Lactose	Mutant 2	Yes	No yellow

5) In **mutant 1**, mRNA and active β-galactosidase **were produced** even when they were grown with **only glucose** — the gene is **always** being expressed.
6) This suggests that mutant 1 has a **faulty *lac* repressor**, e.g. in the **absence** of lactose the repressor **isn't able** to bind DNA, so transcription **can** occur and mRNA and active β-galactosidase **are produced**.
7) In **mutant 2**, **mRNA** is produced but **active β-galactosidase isn't** when **lactose** is present — the **gene** is being **transcribed** but it **isn't** producing **active** β-galactosidase.
8) This suggests mutant 2 is producing **faulty β-galactosidase**, e.g. because a **mutation** has affected its active site.

Warm-Up Questions

PRACTICE QUESTIONS

Q1 What is a transcription factor?

Q2 Explain how repressors stop transcription from happening.

Q3 What is RNAi?

Q4 How does RNAi involving siRNA work?

Q5 How does RNAi involving miRNA work?

Exam Question

Tube	Medium	Bacteria	Full length mRNA	Protein
1	+ Oestrogen	Normal	Yes	Active
2	– Oestrogen	Normal	No	No
3	+ Oestrogen	Mutant	No	No
4	– Oestrogen	Mutant	No	No

Q1 An experiment was carried out to investigate gene expression of the Chi protein in genetically engineered bacteria. A mutant bacterium was isolated and analysed to look for mRNA coding for Chi, and active Chi protein production. The results are shown in the table above:

a) What do the results of tubes 1 and 2 suggest about the control of gene expression? Explain your answer. [2 marks]

b) What do the results of tubes 3 and 4 suggest could be wrong with the mutant? Explain your answer. [3 marks]

c) If an siRNA complementary to the Chi gene was added to tube 1, what would you expect the results to be? Explain your answer. [3 marks]

Transcription Factor — not quite as eXciting as that other factor programme...

If it was a competition, oestrogen would totally win — it's very jazzy and awfully controlling. Flexible too — sometimes it helps to activate and other times it helps to repress. Although I'm not sure it can hold a note or wiggle in time to music. Make sure that you understand everything on these pages. Transcription factors are pretty important molecules.

Epigenetic Control of Gene Expression

Epigenetic changes are another way of controlling gene expression. If you thought transcription factors and the like were clever, then you're in for a real treat with this lot. Prepare to have your mind well and truly blown...

Epigenetic Control Can Determine Whether or Not a Gene is Expressed

1) In **eukaryotes**, **epigenetic control** can determine whether a **gene** is **switched on** or **off** — i.e. whether the gene is **expressed** (transcribed and translated) or not.
2) It works through the **attachment** or **removal** of **chemical groups** (known as **epigenetic marks**) to or from **DNA** or **histone proteins** (see below).
3) These epigenetic marks **don't alter** the **base sequence** of **DNA**.
4) Instead, they **alter** how **easy** it is for the **enzymes** and other proteins needed for **transcription** to **interact** with and **transcribe** the **DNA**.
5) Epigenetic changes to gene expression play a **role** in lots of **normal cellular processes** and can **also occur** in **response** to **changes** in the **environment** — e.g. pollution and availability of food.

Epigenetic Changes Can be Inherited by Offspring

1) Organisms **inherit** their **DNA base sequence** from their **parents**.
2) Most **epigenetic marks** on the DNA are **removed** between generations, but **some escape** the **removal process** and are **passed on** to **offspring**.
3) This means that the expression of some genes in the **offspring** can be **affected** by **environmental changes** that affected their **parents or grandparents**.
4) For example, epigenetic changes in some **plants** in **response** to **drought** have been shown to be **passed on** to later generations.

This epigenetic change, caused by environmental exposure to too many cheesy CGP jokes, has been passed on to three generations so far.

Increased Methylation of DNA Switches a Gene Off

One method of **epigenetic control** is **methylation** of **DNA**:

1) This is when a **methyl group** (an example of an **epigenetic mark**) is **attached** to the **DNA** coding for a **gene**.
2) The group always attaches at a **CpG site**, which is where a **cytosine** and **guanine** base are **next to** each other in the DNA (linked by a **phosphodiester bond**).
3) **Increased** methylation **changes** the **DNA structure** so that the **transcriptional** machinery (enzymes, proteins etc.) **can't interact** with the gene — so the gene is **not expressed** (i.e. it's **switched off**).

A methyl group is a $-CH_3$ group.

Decreased Acetylation of Histones Can also Switch Genes Off

Histones are **proteins** that DNA **wraps around** to form **chromatin**, which makes up **chromosomes**. Chromatin can be **highly** condensed or **less** condensed. How **condensed** it is affects the **accessibility** of the **DNA** and whether or not it can be **transcribed**.

1) Histones can be **epigenetically modified** by the **addition** or **removal** of **acetyl groups** (which are another example of an **epigenetic mark**).
2) When histones are **acetylated**, the chromatin is **less condensed**. This means that the transcriptional machinery **can access** the DNA, allowing genes to be **transcribed**.
3) When **acetyl groups** are **removed** from the histones, the chromatin becomes **highly condensed** and genes in the DNA **can't** be **transcribed** because the transcriptional machinery **can't** physically **access** them.
4) **Histone deacetylase** (**HDAC**) enzymes are responsible for **removing** the **acetyl groups**.

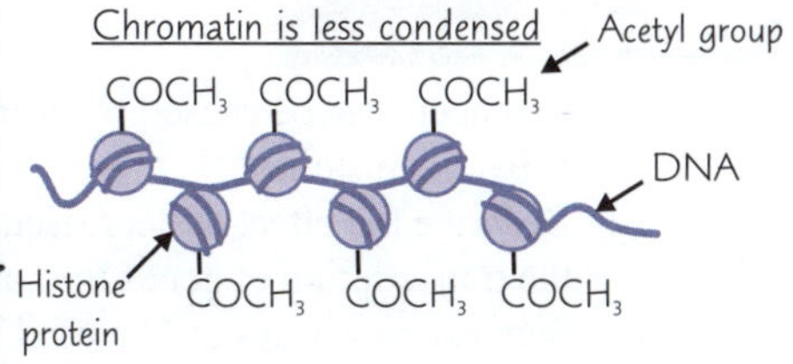

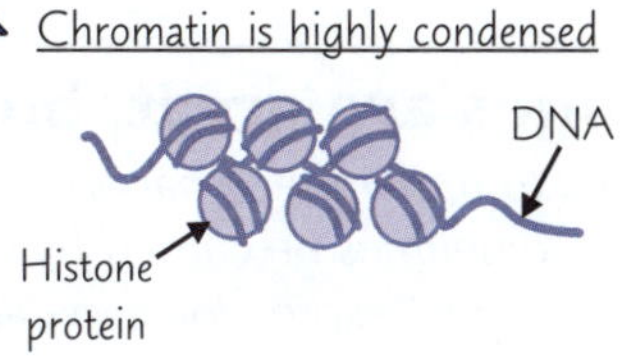

Epigenetic Control of Gene Expression

Epigenetics Can Lead to the **Development** of **Disease**

You've already seen on page 87 how **epigenetics** can play a role in the **development** of **disease**, with the fact that **abnormal methylation** of **tumour suppressor genes** and **oncogenes** can cause **cancer**. However, the role of epigenetics in disease **doesn't** stop there. It can play a role in the development of many other diseases, including Fragile X syndrome, Angelman's syndrome and Prader-Willi syndrome.

Example: Fragile-X syndrome

1) **Fragile-X syndrome** is a genetic disorder that can cause symptoms such as **learning** and **behavioural difficulties**, as well as **characteristic physical features**.
2) It's caused by a heritable **duplication mutation** (see page 84) in a **gene** on the **X chromosome**, called **FMR1**. The mutation results in the **short DNA sequence CGG** being **repeated** many **more** times than usual.
3) These repeats mean that there are lots more **CpG sites** (see previous page) in the gene than usual. More CpG sites result in **increased methylation** of the gene, which **switches it off**.
4) Because the gene is **switched off**, the **protein** that it codes for **isn't produced**. It's the **lack** of this **protein** that causes the **symptoms** of the disease.

Drugs May be Able to **Treat Diseases** Caused by **Epigenetic Changes**

1) **Epigenetic changes** are **reversible**, which makes them **good targets** for new **drugs** to combat **diseases** they cause.
2) These drugs are designed to **counteract** the epigenetic changes that **cause** the **diseases**.
3) For example, **increased methylation** is an epigenetic change that can lead to a **gene** being **switched off**. Drugs that **stop DNA methylation** can sometimes be used to treat diseases caused in this way. For example, the drug **azacitidine** is used in **chemotherapy** for types of **cancer** that are caused by **increased methylation** of **tumour suppressor genes**.
4) **Decreased acetylation** of **histones** can also lead to **genes** being **switched off**. **HDAC inhibitor drugs**, e.g. **romidepsin**, can be used to **treat diseases** that are caused in this way — including some types of **cancer**. These drugs work by **inhibiting** the **activity** of **histone deacetylase** (**HDAC**) **enzymes**, which are responsible for **removing** the **acetyl groups** from the **histones**. Without the activity of HDAC enzymes, the genes **remain acetylated** and the proteins they code for **can be transcribed**.
5) The problem with developing drugs to counteract epigenetic changes is that these changes take place normally in a lot of cells, so it's important to make sure the drugs are as **specific** as possible. E.g. drugs used in **cancer therapies** can be designed to only **target dividing cells** to avoid damaging normal body cells.

Warm-Up Questions

PRACTICE QUESTIONS

Q1 What is epigenetic control?

Q2 What are epigenetic marks?

Q3 How can methylation of DNA affect gene expression?

Exam Question

Q1 Some cancers can be caused by decreased acetylation of histones associated with genes related to cell division.

a) What are histones? [1 mark]

b) Describe the effect of decreased acetylation of histones on the transcription of genes they are associated with. [3 marks]

c) Suggest how drugs can be used to treat cancers caused in this way. [3 marks]

Histones are great, but hisrhythm is way off...

You need to remember what epigenetic control is all about. It's a method for determining whether a gene is transcribed that can sometimes be caused by environmental changes and can be inherited by your offspring. Sometimes, epigenetic changes to gene expression can cause nasty diseases but, in some cases, drugs can be created to cancel them out.

Evaluating Data on Phenotypes

We're finally at the end of the section... and what a whopper it was — but before you rush off to the next one, there's a little bit to learn about the relative influences of genetics and the environment on phenotype...

You Might Have to **Evaluate Data** About **Influences** on **Phenotypes**

The **phenotype** (characteristics) of an organism is the result of the organism's **genotype** and the **interaction** of its genotype with the **environment** (see page 58). It's not always clear **how much** a phenotype is influenced by genes and how much it's influenced by the environment. Let's take a look at these examples:

Example 1 — Overeating

1) **Overeating** was thought to be caused only by environmental factors, like an **increased availability of food** in developed countries.
2) It was later discovered that food consumption **increases** brain **dopamine** levels in animals.
3) Once enough dopamine was released, people would **stop** eating.
4) Researchers discovered that people with one particular **allele** had **30% fewer** dopamine receptors.
5) They found that people with this particular allele were **more likely** to overeat — they wouldn't stop eating when dopamine levels increased.
6) Based on this evidence, scientists now think that overeating has **both genetic** and **environmental** causes.

Example 2 — Antioxidants

1) Many foods in our diet contain **antioxidants** — compounds that are thought to play a role in **preventing chronic diseases**.
2) Foods such as **berries** contain **high levels** of antioxidants.
3) Scientists thought that the berries produced by different **species** of plant contained **different levels** of antioxidants because of **genetic factors**.
4) But experiments that were carried out to see if **environmental** conditions affected antioxidant levels found that environmental conditions caused a great deal of **variation**.
5) Scientists now believe that antioxidant levels in berries are due to **both genetic** and **environmental** factors.

In the exam, you might have to **evaluate data** on the relative influences of genes and the environment on phenotype. This data may come from twin studies...

Twin Studies Can Help to **Determine** Influences on **Phenotype**

Studies of **identical twins** are extremely **useful** when trying to **determine** what's due to **environmental factors** and what's due to **genetic factors**.

These twins are **genetically identical**, so any **differences** in **phenotype** must be entirely due to **environmental factors**. If a characteristic is **very similar** in **identical twins**, **genetics** probably plays a **more important** role. But if a characteristic is **different between** the **twins**, the **environment** must have a **larger influence**.

Twin studies can be used to determine whether a shared bad taste in hats is genetic or just down to buy one get one free offers.

Data that comes from twin studies involving a large sample size (i.e. lots of pairs of twins) is better for drawing valid conclusions than data based on a small sample size. That's because a large sample size is more representative of the population.

Warm-Up Questions

Q1 Give an example of a characteristic that varies due to both genetic and environmental factors.

PRACTICE QUESTIONS

Exam Question

Q1 Twin studies have found that stuttering (a speech disorder) of both twins is more common in identical twins than in non-identical twins.
What do these findings suggest about the influence of genetic and environmental factors on stuttering? [1 mark]

I just don't think there's anything funny about this page...

Like I say, it's been a pretty heavy section. Evaluate means look at both sides of argument and give an overall judgement about something. The important thing is to look at the data properly — don't just skim over it and leap to a conclusion.

Genome Projects and Making DNA Fragments

Gene technologies are seriously amazing. From sequencing the entire human genome to chopping out bits of DNA to insert into other organisms, you never know what scientists will get up to next.

Sequencing Projects Have Read **Entire Genomes**

1) A genome is the **entire set** of DNA, including all the genes in an organism.
2) **Improvements** in **technology** have allowed us to **sequence** the **genomes** of a variety of organisms, from bacteria to humans.
3) Gene sequencing **methods** only work on fragments of DNA. So if you want to sequence the **entire genome** of an organism, you need to chop it up into **smaller pieces** first. The smaller pieces are **sequenced** and then **put back in order** to give the sequence of the whole genome.
4) The **Human Genome Project**, which was completed in 2003, mapped the **entire sequence** of the **human genome** for the first time.

Sequencing the **Genome** of **Simple Organisms** Helps Identify their **Proteins**

1) The **proteome** of an organism is all the **proteins** that are made by it.
2) You might remember from Topic 4 that while some parts of the genome code for **specific proteins**, some parts don't code for anything at all (the DNA is **non-coding**).
3) **Simple** organisms, such as **bacteria**, don't have much non-coding DNA.
4) This means it is relatively **easy** to **determine** their **proteome** from the DNA sequence of their **genome**.
5) This can be useful in **medical research** and **development**. For example, identifying the **protein antigens** on the surface of **disease-causing bacteria** and **viruses** can help in the development of **vaccines** to prevent the disease.

Example: ***N. meningitidis*** **group B bacteria** cause **meningitis B**. **Sequencing** the **genome** of these bacteria helped researchers identify **antigens** for use in developing a **vaccine** against the disease.

Remember: vaccines contain antigens that cause your body to produce memory cells. If you're later infected by a pathogen with the same antigens, your memory cells will quickly recognise it and divide to produce antibodies against it — so you don't get sick.

It's **Harder** to **Translate** the **Genome** of **Complex Organisms**

1) More **complex organisms** contain **large sections** of **non-coding** DNA.
2) They also contain complex **regulatory genes**, which determine when the genes that code for particular proteins should be **switched on** and **off**.
3) This makes it **more difficult** to **translate** their **genome** into their **proteome**, because it's hard to find the bits that code for proteins among the non-coding and regulatory DNA.
4) However, work is being done on the **human proteome**. The codes for more than 30 000 human proteins have been identified so far.

Yes, Sofia was quite sure she didn't need any more sequin-cing.

Sequencing Methods are **Continuously Updated**

1) In the **past**, many sequencing methods were **labour-intensive**, **expensive** and could only be done on a **small scale**.
2) Now these techniques are often **automated**, more **cost-effective** and can be done on a **large scale**.
3) For example, **pyrosequencing** is a **recently developed** technique that can sequence around **400 million bases** in a ten hour period (which is **super fast** compared to older techniques).
4) With **newer**, **faster** techniques such as pyrosequencing available, scientists can now sequence **whole genomes** much more quickly.

Genome Projects and Making DNA Fragments

Recombinant DNA Technology Involves Transferring Fragments of DNA

1) **Recombinant DNA technology** involves **transferring** a **fragment** of **DNA** from **one organism** to **another**.
2) Because the genetic code is **universal** (the same DNA base triplets code for the same amino acids in **all living things**), and because **transcription** and **translation mechanisms** are pretty similar too, the transferred DNA can be used to produce a **protein** in the cells of the **recipient organism**. The recipient and donor organisms don't even have to be from the same species. This can be pretty useful — see page 105. Organisms that contain **transferred DNA** are known as **transgenic organisms**.

DNA Fragments Can Be Made in Different Ways

In order to **transfer** a **gene** from one organism to another, you first need to get a **DNA fragment** containing the gene you're **interested** in (the **target gene**). There are **three ways** that DNA fragments can be produced:

1 Using Reverse Transcriptase

1) Most **cells** only contain **two copies** of each gene, making it **difficult** to obtain a DNA fragment containing the target gene. But they can contain **many mRNA** molecules which are complementary to the gene, so mRNA is often **easier** to obtain.
2) The mRNA molecules can be used as **templates** to **make lots of DNA**. The **enzyme**, **reverse transcriptase**, **makes DNA** from an RNA template. The DNA produced is called **complementary DNA** (**cDNA**).
3) For example, **pancreatic cells** produce the protein **insulin**. They have loads of mRNA molecules complementary to the **insulin gene**, but only **two copies** of the gene **itself**. So reverse transcriptase could be used to **make cDNA** from the **insulin mRNA**.
4) To do this, **mRNA** is first isolated from cells. Then it's **mixed** with **free DNA nucleotides** and **reverse transcriptase**. The reverse transcriptase uses the mRNA as a **template** to synthesise a **new strand** of cDNA.

You should remember from Topic 4 that DNA is copied into mRNA during transcription.

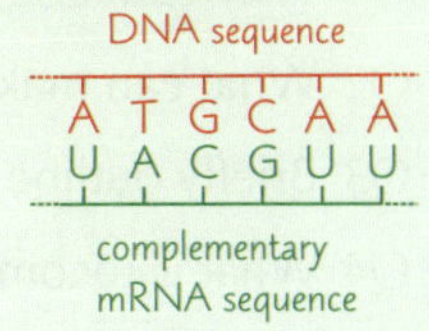

2 Using Restriction Endonuclease Enzymes

1) Some sections of DNA have **palindromic** sequences of **nucleotides**. These sequences consist of **antiparallel base pairs** (base pairs that read the **same** in **opposite directions**).

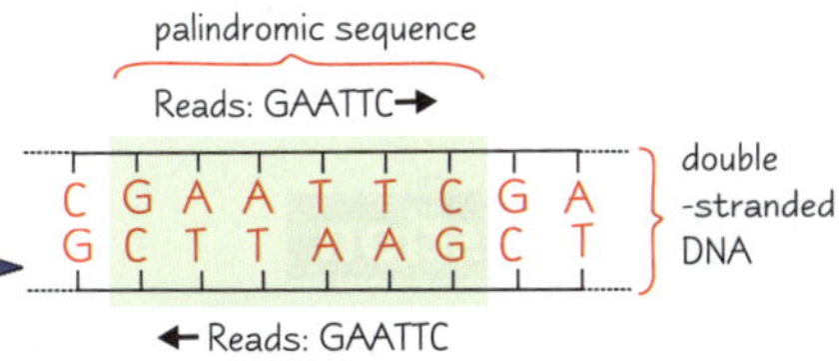

2) **Restriction endonucleases** are enzymes that **recognise specific** palindromic sequences (known as **recognition sequences**) and **cut** (**digest**) the DNA at these places.
3) Different restriction endonucleases cut at **different specific** recognition sequences, because the **shape** of the recognition sequence is **complementary** to the enzyme's **active site**. E.g. the restriction endonuclease *Eco*RI cuts at GAATTC, but *Hind*III cuts at AAGCTT.
4) If recognition sequences are present at **either side** of the DNA fragment you want, you can use restriction endonucleases to **separate** it from the rest of the DNA.
5) The DNA sample is **incubated** with the specific restriction endonuclease, which **cuts** the DNA fragment out via a **hydrolysis reaction**.
6) Sometimes the cut leaves **sticky ends** — **small tails** of **unpaired bases** at **each end** of the fragment. Sticky ends can be used to **bind** (**anneal**) the DNA fragment to another piece of DNA that has sticky ends with **complementary sequences** (there's more about this on p. 103).

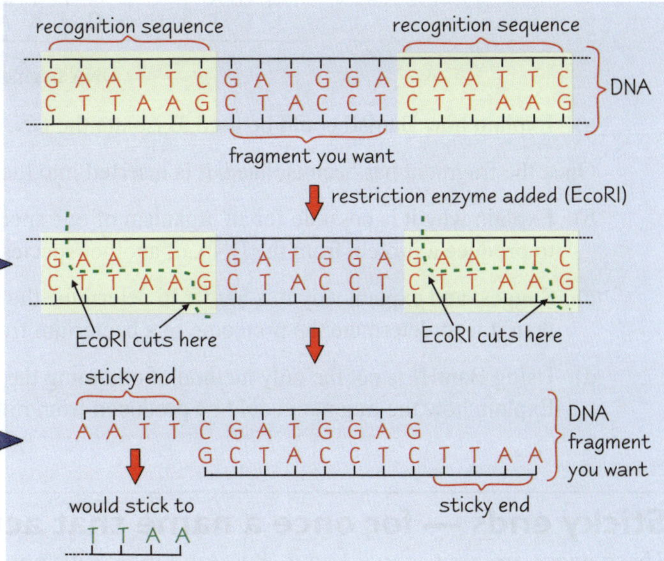

Genome Projects and Making DNA Fragments

3) Using a 'Gene Machine'

1) More **recently**, technology has been developed so that fragments of DNA can be **synthesised** from **scratch**, without the need for a **pre-existing DNA template**.
2) Instead, a **database** contains the necessary information to produce the **DNA fragment**.
3) This means that the DNA sequence does not have to **exist naturally** — **any sequence** can be made.
4) Here's how it's done:

- The **sequence** that is required is **designed** (if one doesn't already exist).
- The first **nucleotide** in the sequence is fixed to some sort of support, e.g. a bead.
- Nucleotides are added **step by step** in the **correct order**, in a cycle of processes that includes adding **protecting groups**. Protecting groups make sure that the nucleotides are **joined** at the **right points**, to prevent **unwanted branching**.
- Short sections of DNA called **oligonucleotides**, roughly 20 nucleotides long, are produced. Once these are complete, they are broken off from the support and all the protecting groups are removed. The oligonucleotides can then be joined together to make **longer DNA fragments**.

A jean machine would be perfect for Rob. He loves his denim.

Warm-Up Questions

PRACTICE QUESTIONS

Q1 Why can it be useful to determine the proteome of a simple organism?

Q2 What can make it difficult to determine the proteome of more complex organisms?

Q3 Briefly outline how genetic sequencing methods have changed over time.

Q4 What is recombinant DNA technology?

Q5 Give three ways a DNA fragment can be produced.

Q6 What is reverse transcriptase?

Q7 What are sticky ends?

Exam Question

Q1 A fragment of DNA (shown below) needs to be isolated from some bacterial DNA.
The restriction endonuclease BamHI recognises the sequence GGATCC and cuts between G and G.

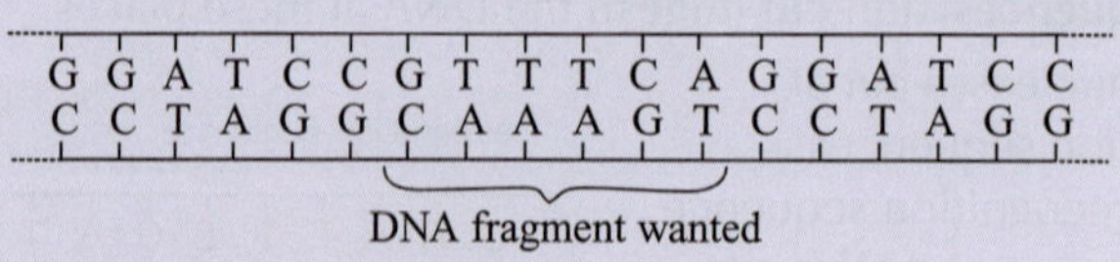

a) Explain how BamHI could be used to isolate the DNA fragment. [2 marks]

Once the fragment has been isolated, it is inserted into the DNA of a plant.

b) Explain why it is possible for an organism of one species to produce a protein from the DNA of another species. [2 marks]

c) Suggest and explain why it is harder to determine the proteome of a plant from its genome than it is to determine the proteome of a bacterium from its genome. [3 marks]

d) Using BamHI is not the only method of obtaining the DNA fragment. Explain how the fragment could be produced from mRNA. [3 marks]

Sticky ends — for once a name that actually makes sense...

These pages are a bit scary I know. But don't worry, it's not as difficult as photosynthesis — you just need to keep going over the steps of the different techniques until they make sense. I know I've said it before, but drawing out the diagrams will help — then you'll know reverse transcriptase and restriction endonucleases like a pro.

Amplifying DNA Fragments

Once you've got your teeny tiny fragment of DNA, you need to amplify it so you've got lots and lots to play with...

***In Vivo* Amplification** Involves **Transforming Host Cells**

Once you've **isolated** your **DNA fragment** (using one of the techniques on pages 101-102) you need to **amplify** it (make lots of copies of it) so you have a **sufficient quantity** to work with. One way of doing this is to use ***in vivo* cloning** — this is where **copies** of the DNA fragment are made **inside** a **living organism**.

Step 1 — The **DNA Fragment** is **Inserted** into a **Vector**

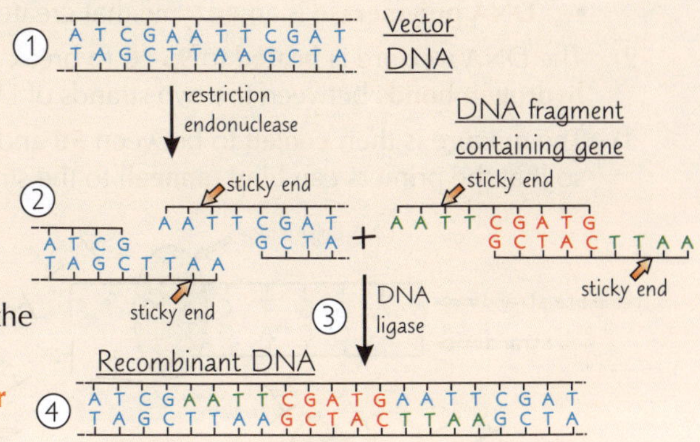

1) The DNA fragment is inserted into vector DNA — a **vector** is something that's used to **transfer DNA** into a **cell**. They can be **plasmids** (**small**, **circular molecules** of DNA in **bacteria**) or **bacteriophages** (**viruses** that **infect** bacteria).
2) The vector DNA is **cut open** using the **same** restriction endonuclease that was used to **isolate** the DNA fragment containing the target gene (see p. 101). So the **sticky ends** of the vector are **complementary** to the sticky ends of the DNA fragment containing the gene.
3) The vector DNA and DNA fragment are **mixed together** with **DNA ligase** (another enzyme). DNA ligase **joins** the sticky ends of the DNA fragment to the sticky ends of the vector DNA. This process is called **ligation**.
4) The new combination of bases in the DNA (vector DNA + DNA fragment) is called **recombinant DNA**.

Step 2 — The **Vector Transfers** the **DNA Fragment** into **Host Cells**

1) The **vector** with the **recombinant DNA** is used to **transfer** the gene into **cells** (called **host** cells).
2) If a **plasmid vector** is used, **host cells** have to be **persuaded** to **take in** the plasmid vector and its DNA.

   E.g. host bacterial cells are placed into ice-cold calcium chloride solution to make their cell walls more permeable. The plasmids are added and the mixture is heat-shocked (heated to around 42 °C for 1-2 minutes), which encourages the cells to take in the plasmids.
3) With a **bacteriophage** vector, the bacteriophage will **infect** the host bacterium by **injecting** its **DNA** into it. The phage DNA (with the target gene in it) then **integrates** into the bacterial DNA.
4) **Host cells** that **take up** the vectors containing the gene of interest are said to be **transformed**.

Step 3 — **Identifying Transformed** Host Cells

Only **around 5%** of host cells will **take up** the vector and its DNA, so it's important to be able to identify which cells have been transformed. **Marker genes** can be used to **identify** the **transformed** cells:

1) **Marker genes** can be inserted into vectors at the **same time** as the gene to be cloned. This means any **transformed host cells** will contain the gene to be cloned **and** the marker gene.
2) Host cells are **grown** on **agar plates**. Each cell **divides** and **replicates** its DNA, creating a **colony** of **cloned cells**. Transformed cells will produce colonies where **all the cells** contain the cloned gene and the marker gene.
3) The marker gene can code for **antibiotic resistance** — host cells are grown on agar plates **containing** the specific **antibiotic**, so **only** transformed cells that have the **marker gene** will **survive** and **grow**. Or it can code for **fluorescence** — when the agar plate is placed under a **UV light only** transformed cells will **fluoresce**.
4) **Identified** transformed cells are allowed to **grow more**, producing **lots** and **lots** of **copies** of the **cloned gene**.

To Produce **Proteins** You Need **Promoter** and **Terminator Regions**

1) If you want the **transformed host cells** to **produce** the **protein** coded for by the **DNA fragment**, you need to make sure that the **vector** contains **specific promoter** and **terminator regions**.
2) **Promoter regions** are **DNA sequences** that tell the enzyme **RNA polymerase** when to **start** producing **mRNA**. **Terminator regions** tell it when to **stop**. Without the **right** promoter region, the DNA fragment **won't** be transcribed by the host cell and a protein **won't** be made.
3) Promoter and terminator regions may be present in the **vector DNA** or they may have to be **added in** along with the **fragment**.

Amplifying DNA Fragments

In Vitro Amplification Uses the **Polymerase Chain Reaction (PCR)**

DNA fragments can also be amplified using ***in vitro*** cloning — this is where copies of the DNA fragments are made **outside** of a living organism using the **polymerase chain reaction** (PCR). PCR can be used to make **millions of copies** of a fragment of DNA in just a few hours. PCR has **several stages** and is **repeated** over and over to make lots of copies:

1) A reaction mixture is set up that contains the **DNA sample**, **free nucleotides**, **primers** and **DNA polymerase**.
   - **Primers** are short pieces of DNA that are **complementary** to the bases at the **start** of the fragment you want.
   - **DNA polymerase** is an **enzyme** that creates new DNA strands.
2) The DNA mixture is **heated** to **95 °C** to break the **hydrogen bonds** between the two strands of DNA.
3) The mixture is then **cooled** to between **50** and **65 °C** so that the primers can **bind** (**anneal**) to the strands.

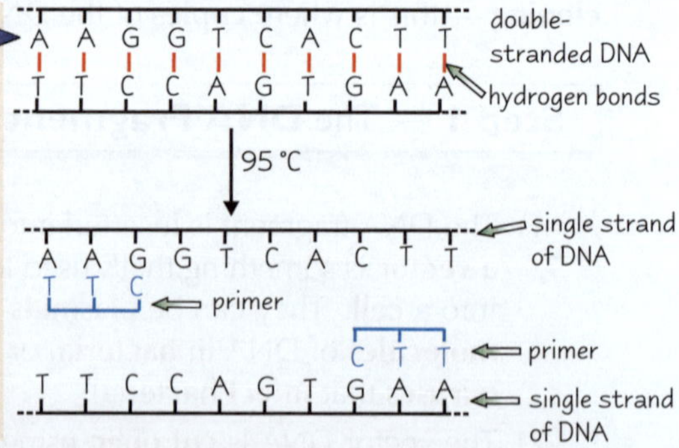

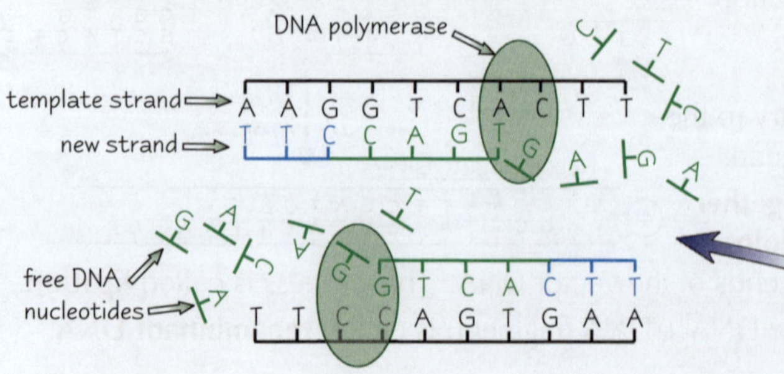

4) The reaction mixture is heated to **72 °C**, so **DNA polymerase** can **work**.
5) The DNA polymerase **lines up** free DNA nucleotides **alongside** each **template strand**. Specific **base pairing** means **new complementary strands** are formed.

6) **Two new copies** of the fragment of DNA are formed and **one cycle** of PCR is **complete**.
7) The cycle starts again, with the mixture being heated to 95 °C and this time **all four strands** (two original and two new) are used as **templates**.
8) Each PCR cycle **doubles** the amount of DNA, e.g. **1st cycle** = 2 × 2 = **4 DNA fragments**, **2nd cycle** = 4 × 2 = **8 DNA fragments**, **3rd cycle** = 8 × 2 = **16 DNA fragments**, and so on.

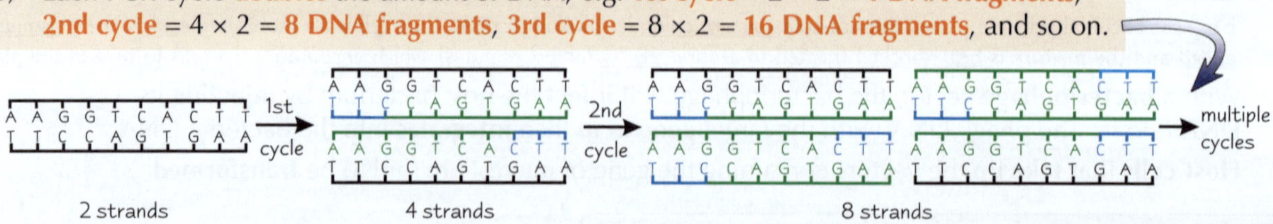

Warm-Up Questions

Q1 In *in vitro* amplification what are vectors used to do?

Q2 What is recombinant DNA?

Q3 What does PCR stand for?

Q4 What is a primer?

PRACTICE QUESTIONS

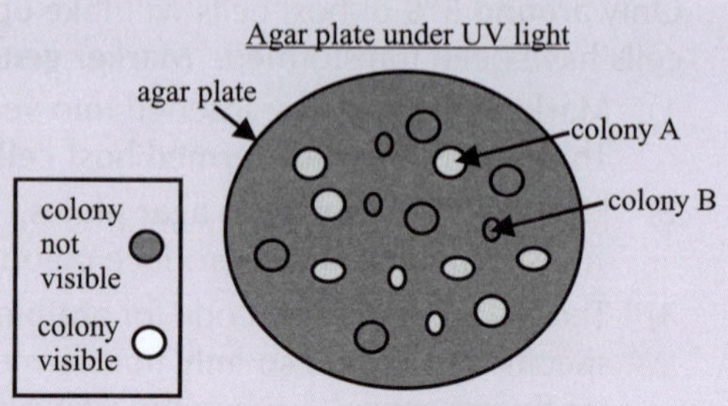

Exam Question

Q1 A scientist amplified a gene by transferring a plasmid containing the target gene and a fluorescent marker gene into some bacterial cells. The cells were grown on an agar plate. The plate was then placed under UV light (see above).

a) Explain why the scientist thinks colony A contains transformed host cells, but colony B doesn't. [2 marks]

b) Explain how the scientist might have inserted the target gene into the plasmid. [3 marks]

If only you could amplify fragments of knowledge — or cake...

Okay, your eyes might have gone funny from seeing so many nucleotides on these pages. But once you've recovered, it's really important to go over these pages as many times as you need to, 'cause examiners love throwing in a few questions on restriction enzymes or PCR. Bless 'em — examiners get excited about the strangest things.

Using Recombinant DNA Technology

Now that you know how to make a DNA fragment and amplify it, it's probably a good time to tell you why you might want to. Don't worry — it's not evil stuff, but I promise to do my evil laugh. Mwah ha hah.

Transformed Organisms Are Made Using Recombinant DNA Technology

Transformed organisms are also known as genetically engineered or genetically modified (GM) organisms.

1) **Microorganisms**, **plants** and **animals** can all be **transformed** using **recombinant DNA technology**. This is called **genetic engineering**.
2) **Transformed microorganisms** can be made using the same technology as ***in vivo*** cloning (see page 103). For example, **foreign DNA** can be **inserted** into **microorganisms** to produce **lots** of **useful protein**, e.g. insulin:

The DNA fragment containing the insulin gene is isolated using a technique from pages 101-102. → The DNA fragment is inserted into a plasmid vector (see page 103). → The plasmid containing the recombinant DNA is transferred into a bacterium. → Transformed bacteria are identified and grown. → The insulin produced from the cloned gene is extracted and purified.

3) **Transformed plants** can also be produced. A gene that codes for a **desirable protein** is inserted into a **plasmid**. The plasmid is added to a **bacterium** and the bacterium is used as a **vector** to get the gene into the **plant cells**. If the right **promoter region** has been added along with the gene (see p. 103), the transformed cells will be able to **produce** the **desired protein**.
4) **Transformed animals** can be produced too — a gene that codes for a **desirable protein** can be inserted into an early animal **embryo** or into the **egg cells** of a female. If the gene is inserted into a **very early embryo**, **all** the **body cells** of the resulting **transformed animal** will end up containing the gene. Inserting it into the **egg cells** means that when the female **reproduces**, **all** the **cells** of her **offspring** will contain the gene.
5) **Promoter regions** that are **only activated** in **specific cell types** can be used to **control** exactly which of an animal's **body cells** the **protein** is produced in. If the protein is only produced in **certain cells**, it can be **harvested** more easily. Producing the protein in the **wrong cells** could also **damage** the organism.

Recombinant DNA Technology Can be Used to Benefit Humans

Transformed organisms (microorganisms, plants and animals) can be used in a variety of ways. You need to be able to **interpret information** about how they are used. Here are some examples:

1 Agriculture

- **Agricultural crops** can be **transformed** so that they give **higher yields** or are **more nutritious**. This means these plants can be used to reduce the risk of **famine** and **malnutrition**. Crops can also be transformed to have **pest resistance**, so that **fewer pesticides** are needed. This **reduces costs** and reduces any **environmental problems** associated with using pesticides.
- For example, ***Golden Rice*** is a variety of **transformed rice**. It contains **one gene** from a **maize plant** and **one gene** from a **soil bacterium**, which together enable the rice to produce **beta-carotene**. The beta-carotene is used by our bodies to produce **vitamin A**. *Golden Rice* is being developed to **reduce vitamin A deficiency** in areas where there's a **shortage** of **dietary vitamin A**, e.g. **south Asia**, **Africa**. Vitamin A deficiency is a big problem in these areas, e.g. up to **500 000 children per year** worldwide go **blind** due to vitamin A deficiency.

2 Industry

- **Industrial processes** often use **biological catalysts** (**enzymes**). These enzymes can be produced from **transformed organisms**, so they can be produced in **large quantities** for **less money**, **reducing costs**.
- For example, **chymosin** (or **rennin**) is an enzyme used in **cheese-making**. It used to be made from **rennet** (a substance produced in the **stomach** of **cows**), but it can now be produced by **transformed organisms**. This means it can be made in **large quantities**, relatively **cheaply** and **without killing** any **cows**, making some cheese suitable for **vegetarians**.

3 Medicine

- Many **drugs** and **vaccines** are produced by transformed organisms, using recombinant DNA technology. They can be made **quickly**, **cheaply** and in **large quantities** using this method.
- For example, **insulin** is used to treat **Type 1 diabetes** and used to come from **animals** (cow, horse or pig pancreases). This insulin **wasn't** human insulin though, so it **didn't work quite as well**. Human insulin is now made from **transformed microorganisms**, using a **cloned human insulin gene** (see above).

Using Recombinant DNA Technology

There are **Concerns** About the Use of **Recombinant DNA Technology...**

There are **ethical**, **financial** and **social issues** associated with the **use** of **recombinant DNA technology**:

1 Agriculture

- **Farmers** might plant only **one type** of transformed crop (this is called **monoculture**). This could make the **whole crop vulnerable** to the **same disease** because the plants are **genetically identical**. Environmentalists are also concerned about monocultures **reducing biodiversity**, as this could damage the environment.
- Some people are concerned about the possibility of '**superweeds**' — weeds that are **resistant** to **herbicides**. These could occur if transformed crops **interbreed** with **wild plants**. There could then be an **uncontrolled spread** of **recombinant DNA**, with **unknown consequences**.
- **Organic farmers** can have their crops **contaminated** by **wind-blown seeds** from nearby **genetically modified** crops. This means they can't **sell** their crop as organic and may **lose** their **income**.

2 Industry

- **Anti-globalisation activists oppose globalisation** (e.g. the **growth** of **large multinational companies** at the **expense** of **smaller ones**). A **few**, **large** biotechnology companies **control** some forms of genetic engineering. As the **use** of this technology **increases**, these companies get **bigger** and **more powerful**. This may **force** smaller companies **out of business**, e.g. by making it **harder** for them to **compete**.
- **Without proper labelling**, some people think they **won't** have a **choice** about whether to consume food made using genetically engineered organisms.
- Some **consumer markets**, such as the EU, won't **import GM foods** and products. This can cause an **economic loss** to **producers** who have traditionally sold to those markets.

3 Medicine

- Companies who **own** genetic engineering technologies may **limit** the **use** of technologies that could be **saving lives**.
- Some people worry this technology could be used **unethically**, e.g. to make **designer babies** (babies that have characteristics **chosen** by their parents). This is currently **illegal** though.

Recombinant DNA technology also creates **ownership issues**. Here are some examples:

- There is some debate about who **owns genetic material** from humans once it has been removed from the body — the **donor** or the **researcher**. Some people argue that the **individual** holds the right to their **own genetic information**, however others argue that **value** is **created** by the **researcher** who uses it to develop a **medicine** or in **diagnosis**.
- A small number of **large corporations** own **patents** to particular **seeds**. They can charge **high prices**, sometimes including a 'technology fee', and can require farmers to **repurchase seeds** each year. If non-GM crops are **contaminated** by GM crops, farmers can be **sued** for breaching the patent law.

...But **Humanitarians** Think it will **Benefit People**

Recombinant DNA technology has **many** potential **humanitarian benefits**:

1) **Agricultural crops** could be produced that help **reduce** the risk of **famine** and **malnutrition**, e.g. **drought-resistant** crops for **drought-prone** areas.
2) **Transformed crops** could be used to produce **useful pharmaceutical products** (e.g. **vaccines**) which could make drugs **available** to **more people**, e.g. in areas where **refrigeration** (usually needed for **storing** vaccines) **isn't available**.
3) **Medicines** could be produced more **cheaply**, so more people can **afford** them.
4) Recombinant DNA technology has the potential to be used in **gene therapy** to **treat human diseases** (see next page).

You need to be able to balance the humanitarian benefits with opposing views from environmentalists and anti-globalisation activists (see above).

Using Recombinant DNA Technology

Gene Therapy Could be Used to **Treat** or **Cure Genetic Disorders** and **Cancer**

Recombinant DNA technology could also be used to **treat human diseases**. This is known as **gene therapy**.

Gene therapy isn't being used widely yet but there is a form of somatic gene therapy available and other treatments are undergoing clinical trials.

How it works:

1) Gene therapy involves **altering** the **defective genes** (mutated alleles) inside cells to treat **genetic disorders** and **cancer**.
2) How you do this depends on whether the disorder is caused by a mutated **dominant allele** or two mutated **recessive alleles** (see page 58):
    - If it's caused by two mutated **recessive** alleles you can **add** a working **dominant allele** to make up for them (you '**supplement**' the faulty ones).
    - If it's caused by a mutated **dominant** allele you can '**silence**' the **dominant allele** (e.g. by sticking a bit of DNA in the middle of the allele so it doesn't work any more).

   Both of these processes involve **inserting** a **DNA fragment** into the person's **original DNA**.

How you get the 'new' allele (DNA) inside the cell:

1) The allele is **inserted into cells** using **vectors** (see page 103) just like in **recombinant DNA technology**.
2) Different **vectors** can be used, e.g. altered **viruses**, **plasmids** or **liposomes** (spheres made of lipid).

There are two types of gene therapy:

1) **Somatic therapy** — this involves **altering** the **alleles** in **body cells**, particularly the cells that are **most affected** by the disorder. For example, **cystic fibrosis** (CF) is a genetic disorder that's very **damaging** to the **respiratory system**, so somatic therapy for CF **targets** the epithelial cells lining the lungs. Somatic therapy doesn't affect the individual's **sex cells** (sperm or eggs) though, so any **offspring** could still **inherit** the disease.
2) **Germ line therapy** — this involves **altering** the **alleles** in the **sex cells**. This means that **every cell** of **any offspring** produced from these cells will be **affected** by the gene therapy and they **won't suffer from the disease**. Germ line therapy in humans is currently **illegal** though.

There are also many **ethical issues** associated with gene therapy. For example, some people are worried that the technology could be used in ways **other** than for **medical treatment**, such as for treating the **cosmetic effects** of **aging**. Other people worry that there's the potential to do **more harm** than good by using the technology (e.g. risk of overexpression of genes — gene produces too much of the missing protein).

Warm-Up Questions

PRACTICE QUESTIONS

Q1 What are transformed organisms?

Q2 Give one financial issue associated with the use of recombinant DNA technology in industry.

Q3 What is gene therapy?

Exam Question

Q1 A large agricultural company isolated a gene from bacteria that may increase the drought resistance of wheat plants.

a) Briefly explain how this gene could be used to make a transformed wheat plant. [3 marks]

b) Suggest how the transformed wheat plants might be beneficial to humans. [2 marks]

c) Suggest why anti-globalisation activists may be against the use of this gene. [1 mark]

Neapolitan — recombinant ice cream...

Ahhh, sitting in the Sun, licking an ice cream, exams all over. That's where you'll be in a few months' time. After revising all this stuff that is. As recombinant DNA technology advances, more questions will pop up about its implications. So it's a good idea to know all sides of the argument — you need to know them for the exam anyway.

Gene Probes and Medical Diagnosis

Being able to manipulate DNA is also really useful for diagnosing medical problems...

You can **Look** for **Alleles** Using **DNA Probes** and **Hybridisation**

1) DNA probes can be used to **locate specific alleles** of **genes** (e.g. on **chromosomes**) or to see if a person's DNA contains a **mutated allele** that causes a **genetic disorder**.
2) **DNA probes** are **short strands** of **DNA**. They have a **specific base sequence** that's **complementary** to the base sequence of part of a **target allele** (the allele you're looking for, e.g. an allele that causes a genetic disorder).

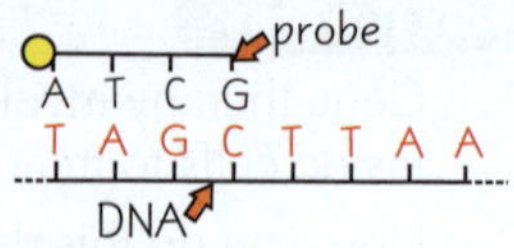

3) This means a DNA probe will **bind** (**hybridise**) to the **target allele** if it's **present** in a **sample** of DNA.
4) A DNA probe also has a **label attached**, so that it can be **detected**. The two most common types of label are a **radioactive** label (detected using **X-ray film**) or a **fluorescent** label (detected using **UV light**).
5) Here's how it's done:

- A **sample** of **DNA** is **digested** into fragments using **restriction enzymes** (see p. 101) and separated using **electrophoresis** (see p. 110).
- The separated DNA fragments are then transferred to a **nylon membrane** and **incubated** with the **fluorescently labelled DNA probe**.
- If the allele is **present**, the **DNA probe** will **hybridise** (**bind**) to it.
- The **membrane** is then **exposed** to **UV light** and if the gene is present there will be a **fluorescent band**. E.g. **sample 3** has a visible band, so this patient has the **allele**.

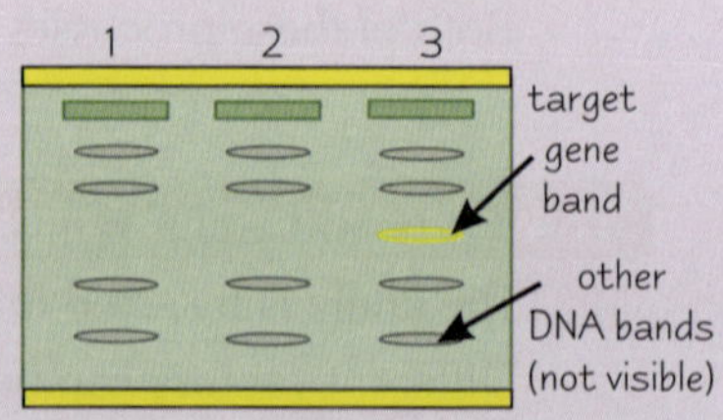

6) Alternatively, the probe can be used as part of a **DNA microarray**, which can screen **lots** of **genes** at the **same time**:

- A **DNA microarray** is a **glass slide** with **microscopic spots** of **different** DNA probes **attached** to it in **rows**.
- A sample of **fluorescently labelled human DNA** is washed over the array.
- If the labelled human DNA **contains** any **DNA sequences** that **match** any of the **probes**, it will **stick** to the array.
- The array is **washed**, to remove any labelled DNA that **hasn't** stuck to it.
- The array is then **visualised** under **UV light** — any **labelled DNA attached** to a probe will **show up** (fluoresce).
- Any spot that fluoresces means that the person's DNA **contains** that specific **allele**. E.g. if the probe is for a mutated allele that causes a **genetic disorder**, this person has the allele.

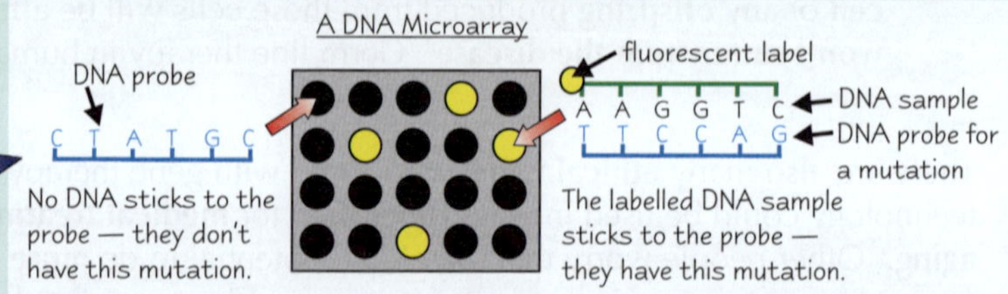

7) To **produce** a DNA probe, you first need to **sequence** the **allele** that you want to screen for (see page 100). You then use **PCR** (see p. 104) to produce **multiple complementary copies** of **part** of the allele — these are the **probes**.

Screening Using **DNA Probes** Has **Lots of Uses**

You need to be able to evaluate information about screening for inherited conditions and people's responses to drugs.

For example, screening can be used to...

1) ...help identify **inherited conditions**. E.g. **Huntington's disease** is an inherited condition that affects the nervous system and does not usually start to display symptoms until a person is aged between 30 and 50. People with a family history of the disease may choose to be **screened** for the **mutated allele** to find out if they have inherited it. The NHS offers to screen **all newborn babies** for the inherited disorder **cystic fibrosis** (which can cause breathing and digestive difficulties) so that **treatment** for the condition can begin as **soon as possible**.
2) ...help determine how a patient will respond to specific drugs (see next page). E.g. **breast cancer** can be caused by a **mutation** in the **HER2 proto-oncogene** and treated with the drug **Herceptin®** (see page 89). Herceptin® is **only effective** against this type of breast cancer because it **targets** a specific receptor. Screening for this particular mutation helps determine whether Herceptin® will be a useful treatment or not.
3) ...help identify **health risks**. E.g. inheriting particular mutated alleles **increases** your **risk** of developing certain **cancers** (although it doesn't make it certain that you'll develop cancer). If a person knows they have these alleles, it might help them to make **choices** that could **reduce the risk** of the disease developing (see next page). However, some people are concerned that **genetic screening** may lead to **discrimination** by **insurance companies** and **employers** if people are known to have a **high risk** of developing a condition.

Gene Probes and Medical Diagnosis

The **Results** of **Screening** can be used for **Genetic Counselling**...

1) **Genetic counselling** is **advising patients** and their **relatives** about the **risks** of **genetic disorders**.
2) It involves **advising** people about **screening** (e.g. looking for mutated alleles if there's a **history** of **cancer**) and **explaining** the **results** of a screening. Screening can help to **identify** if someone is the **carrier** of a mutated allele, the **type** of **mutated allele** they're carrying (indicating the type of genetic disorder or cancer) and the **most effective treatment**.
3) If the results of a screening are **positive** (an individual **has** the mutation) then genetic counselling is used to advise the patient on the **options** of **prevention** or **treatment** available. Here are two examples:

EXAMPLE 1: A **woman** with a family history of **breast cancer** may have **genetic counselling** to help her **decide** whether or not to be **screened** for **known mutations** that can lead to breast cancer, e.g. a mutation in the BRCA1 **tumour suppressor gene** (see p. 89). If she is screened and the result is **positive**, genetic counsellors might explain that a woman with the mutated BRCA1 gene has a **50 to 85%** chance of developing **breast cancer** in her lifetime. Counselling could also help the woman to **decide** if, for example, she wants to take **surgical steps** to **reduce the risk** of breast cancer developing (by having a **mastectomy**).

EXAMPLE 2: **Sickle-cell anaemia** is a **recessive** genetic disorder caused by a **mutation** in the **haemoglobin gene**. A couple who are **both carriers** of the **sickle-cell allele** may **like** to have **kids**. They may undergo genetic counselling to help them **understand** their **chances** of having a child with sickle-cell anaemia (**one in four**). Genetic counselling also provides **unbiased advice** on the possibility of having **IVF** and **screening** their **embryos** for the allele, so embryos **without the mutation** are **implanted** in the womb. It could also provide information on the **help** and **drugs** available if they have a child with sickle-cell anaemia.

A carrier is a person carrying an allele that is not expressed in their phenotype but that can be passed on to offspring — see page 58.

...and in **Personalised Medicine**

1) Your **genes** determine how your body **responds** to certain **drugs**. **Different people** respond to the **same drug** in **different ways** — which makes certain drugs **more effective** for **some people** than others. This is where **personalised medicines** come in.
2) Personalised medicines are medicines that are **tailored** to an **individual's DNA**. The theory is that if doctors have your **genetic information**, they can use it to **predict** how you will respond to different drugs and only prescribe the ones that will be **most effective** for you.

Warm-Up Questions

PRACTICE QUESTIONS

Q1 What are DNA probes?

Q2 Give three situations where screening for mutated genes may be useful.

Q3 What is personalised medicine?

Exam Questions

Q1 a) Briefly describe how a DNA probe for a clinically important allele can be produced. [2 marks]

b) Describe how you could screen a person for this allele and many other alleles at the same time. [4 marks]

Q2 A hospital patient has colon cancer. A drug called Cetuximab is used to treat colon cancer caused by a mutation in the KRAS proto-oncogene. The patient is screened and tests negative for the KRAS oncogene.

a) Why is it unlikely that the patient will be treated with Cetuximab? [1 mark]

b) Suggest why the patient will undergo genetic counselling. [2 marks]

DNA probes — don't worry, the DNA doesn't feel a thing...

All of the techniques you've learnt earlier in this section (making and amplifying DNA fragments, PCR) come together nicely in this medical diagnosis stuff — it's good to know that what you've learnt has a point to it.

Genetic Fingerprinting

We've been able to identify people from their fingerprints for over 100 years, but now we can use their DNA instead.

Genomes Contain Non-Coding **Variable Number Tandem Repeats**

1) **Not all** of an organism's **genome** (all the genetic material in an organism) **codes** for **proteins**.
2) Some of the genome consists of **variable number tandem repeats** (**VNTRs**) — base sequences that **don't** code for proteins and **repeat** next to each other over and over (sometimes thousands of times), e.g. CATGCATGCATGCATG is a repeat of the non-coding base sequence CATG.
3) The **number of times** these sequences are **repeated differs** from person to person, so the **length** of these sequences in nucleotides differs too. E.g. a **four** nucleotide sequence might be repeated **12 times** in one person = **48 nucleotides** (12 × 4), but repeated **16 times** in another person = **64 nucleotides** (16 × 4).
4) The repeated sequences occur in **lots of places** in the **genome**. The **number** of times a **sequence is repeated** (and so the number of nucleotides) at **different places** in their genome can be **compared** between **individuals** — this is called **genetic fingerprinting**.
5) The **probability** of **two individuals** having the **same** genetic fingerprint is **very low** because the **chance** of **two individuals** having the **same number** of VNTRs at **each place** they're found in DNA is **very low**.

Electrophoresis Separates DNA Fragments to Make a **Genetic Fingerprint**

So **genetic fingerprints** can be **compared** between **different individuals**. Now you need to know how one is **made**:

1) A **sample** of **DNA** is obtained, e.g. from a person's **blood**, **saliva**, etc.
2) **PCR** (see page 104) is used to make **many copies** of the **areas** of DNA that contain the VNTRs — **primers** are used that bind to **either side** of these **repeats** and so the **whole** repeat is amplified.
3) You end up with **DNA fragments** where the **length** (in nucleotides) corresponds to the **number of repeats** the person has at each specific position, e.g. one person may have 80 nucleotides, another person 120.
4) A **fluorescent tag** is added to all the DNA fragments so they can be viewed under **UV light**.
5) The DNA fragments undergo **electrophoresis**:
   - The DNA mixture is placed into a **well** in a slab of **gel** and covered in a **buffer solution** that **conducts electricity**.
   - An **electrical current** is passed through the gel — DNA fragments are **negatively charged**, so they **move towards** the **positive electrode** at the far end of the gel.
   - **Small** DNA fragments move **faster** and **travel further** through the gel, so the DNA fragments **separate** according to **size**.
6) The DNA fragments are viewed as **bands** under **UV light** — this is the **genetic fingerprint**.
7) Two genetic fingerprints can be **compared** — e.g. if both fingerprints have a band at the **same location** on the **gel** it means they have the **same number** of **nucleotides** and so the same number of **VNTRs** at that place — it's a **match**.

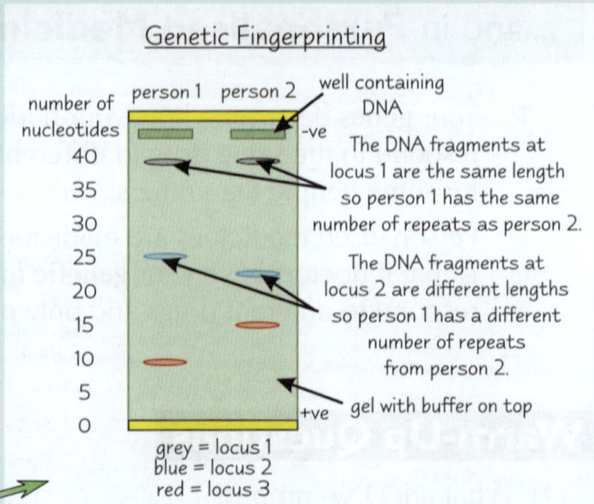

Genetic Fingerprinting is Used to **Determine Relationships** and **Variability**

Genetic fingerprinting has **many uses**, which include:

- **Determining genetic relationships** — We **inherit** VNTR base sequences from our **parents**. Roughly **half** of the sequences come from **each parent**. This means the **more bands** on a genetic fingerprint that match, the more **closely related** (**genetically similar**) two people are. E.g. **paternity tests** are used to determine the **biological father** of a child by comparing genetic fingerprints. If lots of bands on the fingerprint **match**, then that person is **most probably** the child's father. The **higher** the **number** of places in the genome compared, the more **accurate** the test result.
- **Determining genetic variability within a population** — The **greater** the **number of bands** that **don't** match on a genetic fingerprint, the more **genetically different** people are. This means you can **compare** the **number of repeats** at **several places** in the genome for a population to find out how **genetically varied** that population is. E.g. the **more** the **number of repeats** varies at **several places**, the **greater** the **genetic variability** within a population.

Genetic Fingerprinting

Genetic Fingerprinting can be Used in Forensic Science...

Forensic scientists use genetic fingerprinting to **compare** samples of **DNA** collected from **crime scenes** (e.g. DNA from **blood**, **semen**, **skin cells**, **saliva**, **hair**, etc.) to samples of DNA from **possible suspects**, which could **link them** to crime scenes.

1) The **DNA** is **isolated** from all the collected samples (from the crime scene and from the suspects).
2) Each sample is **replicated** using **PCR** (see p. 104).
3) The **PCR products** are run on an **electrophoresis gel** and the genetic fingerprints produced are **compared** to see if any match.
4) If the samples match, it **links** a **person** to the **crime scene**. E.g. this gel shows that the genetic fingerprint from **suspect C matches** that from the crime scene, **linking** them to the crime scene. All five bands match, so suspect C has the **same number** of repeats (nucleotides) at **five** different places.

Example — Genetic Fingerprints

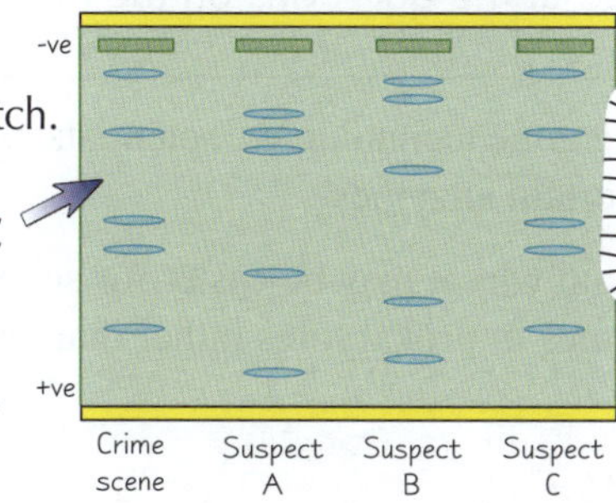

PCR amplifies the DNA, so enough is produced for it to be seen on the gel.

...Medical Diagnosis...

- In medical diagnosis, a genetic fingerprint can refer to a **unique pattern** of **several alleles**.
- It can be used to **diagnose genetic disorders** and **cancer**. It's useful when the **specific** mutation **isn't** known or where **several mutations** could have caused the disorder, because it identifies a **broader**, **altered** genetic pattern.

  EXAMPLE 1: **Preimplantation genetic haplotyping** (**PGH**) **screens embryos** created by **IVF** for genetic disorders **before** they're **implanted** into the uterus. The **faulty regions** of the **parents' DNA** are used to produce **genetic fingerprints**, which are **compared** to the genetic fingerprint of the **embryo**. If the fingerprints **match**, the embryo has **inherited** the **disorder**. It can be used to screen for **cystic fibrosis**, **Huntington's disease**, etc.

  EXAMPLE 2: Genetic fingerprinting can be used to **diagnose sarcomas** (types of **tumour**). Conventional methods of identifying a tumour (e.g. biopsies) only show the **physical differences** between tumours. Now the **genetic fingerprint** of a known sarcoma (e.g. the **different mutated alleles**) can be **compared** to the genetic fingerprint of a **patient's tumour**. If there's a **match**, the sarcoma can be specifically **diagnosed** and the **treatment** can be targeted to that specific type (see page 89).

A specific mutation can be found using DNA probes and sequencing (see p. 108).

...and Animal and Plant Breeding

Genetic fingerprinting can be used on **animals** and **plants** to **prevent inbreeding**, which **decreases** the **gene pool** (the number of **different alleles** in a population, see p. 66). Inbreeding can lead to an **increased risk** of **genetic disorders**, leading to **health**, **productivity** and **reproductive problems**. Genetic fingerprinting can be used to **identify** how **closely-related** individuals are — the **more closely-related** two individuals are, the **more similar** their genetic fingerprint will be (e.g. **more bands** will **match**). The **least related** individuals will be **bred together**.

Warm-Up Questions

Q1 Why are two people unlikely to have the same genetic fingerprint?

Q2 In gel electrophoresis, which electrode do DNA fragments move towards?

Q3 Why might genetic fingerprinting be used in forensic science?

Exam Question

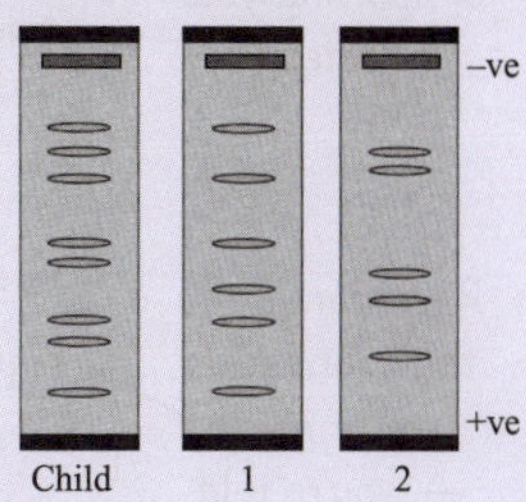

Q1 The diagram shows three genetic fingerprints — one from a child and two from possible fathers.

a) Explain how PCR enables genetic fingerprinting to be carried out. [3 marks]

b) Which genetic fingerprint is most likely to be from the child's father? Explain your answer. [1 mark]

c) Give two more uses of genetic fingerprint technology. [1 mark]

Fingerprinting — in primary school it involved lots of paint and paper...

Who would have thought that tiny pieces of DNA on a gel would be that important? Well, they are and you need to know all about them. Make sure you know the theory behind fingerprinting as well as its applications. And remember, it's very unlikely that two people will have the same genetic fingerprint (except identical twins that is).

Extra Exam Practice

Wow — Topic 8 was no picnic, that's for sure. There were lots of new facts and processes to remember. If you're like me you might understand everything perfectly while you're learning it, but find it all gets a bit jumbled up when you try to recall it. Well, there's one sure-fire way to check everything's unjumbled before exam-day — do some practice exam questions to see how much you know.

- Have a look at this example of how to answer a tricky exam question.
- Then check how much you've understood from Topic 8 by having a go at the questions on the next page.

Think you've done all the practice exam questions? Not quite — there are synoptic questions covering the whole course on pages 123-127.

1 Using gene technologies, scientists have successfully identified the base sequences that code for certain enzymes.

1.1 Explain why it may be easier for scientists to identify the base sequences that code for specific enzymes in prokaryotes rather than in humans.

(3 marks)

One example of a human enzyme that scientists have identified the base sequence of is cytochrome P450 2C19. This enzyme is needed to metabolise the drug clopidogrel, which is often prescribed to patients with cardiovascular disease.
Some doctors will only prescribe clopidogrel to certain patients as part of a personalised medicine scheme, as some people have a version of cytochrome P450 2C19 with an altered sequence of amino acids.

1.2 Using this information, explain why a doctor may choose not to prescribe clopidogrel to a patient with cardiovascular disease.

(2 marks)

1.3 Explain how a doctor could determine whether or not a patient with cardiovascular disease should be prescribed clopidogrel as part of a personalised medicine scheme.

(2 marks)

You should know from earlier in the course that prokaryotes are very simple organisms — once you've worked that out, it's just a case of applying your knowledge.

First point out the relevant differences between the genomes of the organisms. Then explain why this makes identifying the base sequences of specific enzymes more difficult in humans.

1.1

Prokaryotes are **simple organisms** so **don't have much non-coding DNA**, whereas humans are more complex organisms, so their genome contains **large sections of non-coding DNA** as well as **complex regulatory genes**. This makes it more difficult to translate the human genome into its proteome and determine the base sequences for specific enzymes, because it's **hard to find the base sequences that code for proteins** among the non-coding and regulatory DNA.

Pay attention to the information you're given — you're told that cytochrome P450 2C19 is an enzyme so think about how an altered amino acid structure could affect this specific protein.

1.2

If the patient has a version of cytochrome P450 2C19 that codes for an altered sequence of amino acids then the **specific shape** of the **enzyme's active site** may be altered, which could mean that the patient would be unable to metabolise clopidogrel efficiently.

Questions about personalised medicine are likely to involve looking at a patient's genes.

The altered sequence of amino acids must be caused by a different sequence of bases in the gene (a different allele).

1.3

The doctor would need to have the patient **genetically screened** to determine whether they had the **allele** for **cytochrome P450 2C19** that coded for the normal sequence of amino acids, or whether they had an **allele** that coded for an **altered sequence**. If they had the allele that coded for the altered sequence, the doctor may choose not to prescribe the drug.

Extra Exam Practice

2 Myelodysplastic syndrome (MDS) is a group of rare blood disorders.

The defect that causes MDS occurs in the nuclei of haematopoietic stem cells. As a result, MDS can affect the numbers of red blood cells, lymphocytes and phagocytes in the blood.

2.1 Suggest whether haematopoietic stem cells are multipotent or unipotent.
Explain your answer.

(1 mark)

2.2 Haematopoietic stem cells divide to form new cells, which can then develop into red blood cells.
Explain what happens in these new cells in order for them to become red blood cells.

(4 marks)

2.3 A possible cure for MDS is a stem cell transplant.
Suggest why this cure involves receiving cells from a donor rather than using the patient's own cells.

(2 marks)

MDS is classed as a cancer because it causes abnormal cell growth, which is caused by disruption to the normal function of tumour suppressor genes.
MDS can be treated with the drug, azacitidine. When incorporated into DNA, metabolised azacitidine binds irreversibly to DNA methyltransferases, which are the enzymes responsible for DNA methylation.

2.4 Explain how azacitidine works as a treatment for MDS.

(4 marks)

2.5 Explain why azacitidine may be described as an 'epigenetic' drug.

(1 mark)

Many studies have been done to investigate a link between smoking and development of MDS.
The results of some of these studies were analysed to produce odds ratios.
Odds ratios quantify the strength of the connection between exposure to a variable and an outcome. In this case, the greater the value of the odds ratio, the greater the odds of MDS occurring as a result of being exposed to the variable.
The odds ratios for some of the studies analysed are shown in **Table 1**.

Table 1

Study	Odds ratio	
	Current smokers	Former smokers
Lv (2011)	1.37	1.48
Ma (2009)	3.17	1.68
Dalamaga (2002)	4.03	2.60
Mele (1994)	1.20	2.70
Strom (2005), male	1.85	1.31

2.6 Some health experts advise people to stop smoking in order to reduce their risk of developing MDS.
Is this advice supported by the data shown in **Table 1**? Explain your answer.

(2 marks)

PRACTICAL SKILLS

Planning an Experiment

As well as doing practical work in class, you can get asked about it in your exams too. Harsh I know.

Before You Start Planning, **Be Clear** on What You're Trying to **Find Out**

Like all scientists, you should start off by making a **prediction** or **hypothesis** — a **specific testable statement**, based on theory, about what will happen in the experiment. You then need to **plan** a good experiment that will provide **evidence to support the prediction** — or help **disprove it**.

A **Good Experiment** Gives **Results** that are...

Precise results are sometimes referred to as reliable results.

1) **Precise** — precise results **don't vary much** from the **mean**. Precision is reduced by **random error** (the unpredictable way in which all measurements vary).
2) **Repeatable and reproducible** — repeatable means that if the same person repeats the experiment using the same methods and equipment, they will get the same results. Reproducible means that if someone different does the experiment, using a slightly different method or piece of equipment, the results will still be the same.
3) **Valid** — valid results **answer** the **original question**. To get valid results you need to **control all the variables** (see below) to make sure you're only testing the thing you want to.
4) **Accurate** — accurate results are **really close** to the true answer. **Human interpretation** of a measurement (e.g. determining a colour change) can **reduce** the accuracy of results.

Here are some things you need to consider when designing a good experiment:

1) **Only one variable should be changed** — Variables are **quantities** that have the **potential to change**, e.g. pH. In an experiment you usually **change one variable** and **measure its effect** on another variable.
   - The variable that you **change** is called the **independent variable**.
   - The variable that you **measure** is called the **dependent variable**.
2) **All the other variables should be controlled** — When you're investigating a variable you need to keep everything else that could affect it **constant**. This means you can be sure that **only** your **independent** variable is **affecting** the thing you're measuring (the dependent variable).
3) **Negative controls should be used** — Negative controls are used to **check** that only the independent variable is affecting the dependent variable. Negative controls **aren't expected** to have **any effect** on the experiment.
4) **The experiment should be repeated at least three times and a mean should be calculated** — this reduces the effect of **random error** on your experiment, which makes your results **more precise**. Doing repeats and getting **similar results** each time also shows that your data is **repeatable** and makes it more likely to be **reproducible**.

<u>EXAMPLE</u>: Investigating the effect of **light intensity** on **rate of photosynthesis** of **Canadian pondweed**.

1) Light intensity is the **independent** variable.
2) Rate of photosynthesis is the **dependent** variable.
3) pH, temperature and the time the pondweed is left should all **stay the same** (and the quantities should be recorded to allow someone else to reproduce the experiment).
4) The experiment should be **repeated** at least **three times** for each light intensity used.
5) A **negative control**, in which the experiment is carried out in the **dark**, should also be used. No photosynthesis should happen with this control.

Examiners love getting you to **comment** on **experimental design** or **suggest improvements** to **methods** — e.g. how a method could be improved to make the results more precise. So make sure you know how to **design** a **good experiment**.

Select Appropriate **Apparatus, Equipment** and **Techniques**

1) When you're **planning** an experiment you need to decide what it is you're going to **measure** and **how often** you're going to take measurements. E.g. if you're investigating the **rate of respiration**, you could either measure the volume of **oxygen used** over time or the volume of **carbon dioxide produced** over time. You could take measurements at, e.g. 30 second intervals or 60 second intervals.
2) Then you need to choose the most **appropriate** apparatus, equipment and techniques for the experiment. E.g.
   - The **measuring apparatus** you use has to be **sensitive** enough to measure the changes you're looking for. For example, if you need to measure **small changes** in **pH**, a **pH meter** (which can measure pH to several decimal places) would be more sensitive than indicator paper.
   - The **technique** you use has to be the most **appropriate** one for your **experiment**. E.g. if you want to measure the concentration of glucose in an unknown solution, using a **colorimeter** in conjunction with **quantitative Benedict's reagent** (see page 51) will help you to get more **accurate results** than simply comparing the colour differences of the solutions by eye.

Planning an Experiment

PRACTICAL SKILLS

You Need to Know How to Use **Apparatus** and **Techniques Correctly**

Examiners could ask you about a **whole range** of different apparatus and techniques. Make sure you know how to use all the instruments and equipment you've come across in class and can carry out all the techniques too. Here are some **examples** of equipment you should be able to use:

- **Measuring cylinders** and **graduated pipettes** — These have a **scale** so you can measure specific **volumes**. Whichever one you use, make sure you read the volume from the **bottom** of the **meniscus** when it's at **eye level**.

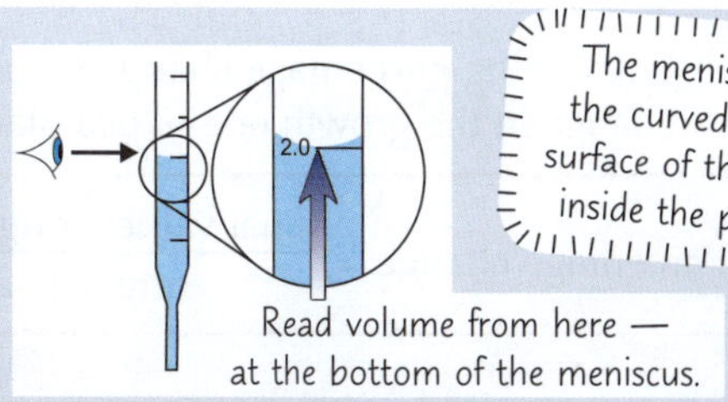

The meniscus is the curved upper surface of the liquid inside the pipette.

- **Water baths** — Make sure you **allow time** for water baths to **heat up** before starting your experiment. Don't forget that your **solutions** will need **time** to get to the **same temperature** as the water before you start the experiment too. Also, remember to **check** the **temperature** of the water bath with a **thermometer** during the investigation to make sure it **doesn't change**.
- **Data logger** — Decide **what** you are **measuring** and what **type** of **data logger** you will need, e.g. temperature, pH. Connect an **external sensor** to the data logger if you need to. Decide **how often** you want the data logger to take readings depending on the **length** of the **process** that you are measuring.

Make sure you know how to do **all** the **practical investigations** described in this book. You should be able to **apply** the techniques described to **different contexts**. For example, page 11 describes how to use a **colorimeter** and a **redox indicator dye** to investigate the rate of **dehydrogenase activity** in **chloroplasts**. You could use a similar technique (i.e. a colorimeter and a redox indicator dye) to investigate the **rate of respiration** in **yeast**.

Risk Assessments Help You to **Work Safely**

1) When you're planning an experiment, you need to carry out a **risk assessment**. To do this, you need to identify:
   - All the **dangers** in the experiment, e.g. any hazardous chemicals, microorganisms or naked flames.
   - **Who** is at **risk** from these dangers.
   - What can be done to **reduce** the **risk**, such as wearing goggles or gloves or working in a fume cupboard.
2) You also need to consider any **ethical issues** in your experiment. For example, if you're using **living animals** (e.g. insects) you must treat them with **respect**. This means **handling them carefully** and keeping them away from **harmful chemicals**, **extreme heat sources** and other things that might cause them **physical discomfort**.

Record Your **Data** in a **Table**

It's a good idea to draw a table to **record** the **results** of your experiment in.

1) When you draw a table, make sure you **include** enough **rows** and **columns** to **record all of the data** you need to. You might also need to include a column for **processing** your data (e.g. working out an average).
2) Make sure each **column** has a **heading** so you know what's going to be recorded where. The **units** should be in the **column heading**, not the table itself.
3) The **independent variable** should be recorded in the **left-hand** column and the **dependent variable** in the **right**.

heading — column — units — row — data

Concentration / mol dm^{-3}	*Absorbance / Absorbance Units (AU)*
0.2	*0.5*
0.4	*0.9*
0.6	*1.3*

Watch Out for **Anomalous Results**

Doing repeats makes it easier to spot anomalous results.

When you look at all the **data** in your **table**, you may notice that you have a result that **doesn't seem to fit in** with the rest at all. These results are called **anomalous results**. You should **investigate** anomalous results — if you can work out what happened (e.g. you measured something totally wrong) you can **ignore** them when **processing** your results. However, you can't just ignore a result because you don't like the look of it.

My best apparatus is the pommel horse...

It's not really, I just like the word pommel. Scientists are rightfully fussy about methods and equipment — I mean if you're going to bother doing an experiment, you should at least make sure it's going to give you results you can trust.

Processing and Presenting Data

PRACTICAL SKILLS

Processing data means taking raw data and doing some calculations with it, to make it more useful.

Processing the Data Helps You to **Interpret** it

You Need to be Able to **Calculate Percentage Change** and **Ratios**

1) Calculating **percentage change** helps to **quantify** how much something has changed, e.g. the percentage change in the growth rate of pea plants when a fertiliser is added. To **calculate** it you use this equation:

$$\text{Percentage change} = \frac{\text{final value} - \text{original value}}{\text{original value}} \times 100$$

A **positive** value shows an **increase** and a **negative** value shows a **decrease**.

E.g. a person's blood glucose concentration before a meal was **4.2 mmol dm^{-3}**. Two hours after a meal it was **6.5 mmol dm^{-3}**. Calculate the percentage change.

$$\text{Percentage change} = \frac{6.5 - 4.2}{4.2} \times 100 = \mathbf{55\%} \text{ (2 s.f.)}$$

So the person's blood glucose concentration was 55% higher after the meal.

2) Ratios can be used to **compare** lots of different types of quantities. E.g. an organism with a **surface area to volume ratio** of **2 : 1** would theoretically have a surface area **twice as large** as its volume.
3) Ratios are usually most useful in their **simplest** (smallest) **form**. To simplify a ratio, **divide each side** by the **same number**. It's in its simplest form when there's nothing left you can divide by. To get a ratio of X : Y in the form **X : 1, divide both sides by Y**. E.g. to get 28 : 34 into the ratio of X : 1, divide both sides by 34. You get 0.82 : 1.

You Need to be Able to **Use Logarithms**

You need to be able to read off a logarithmic scale on a graph.

1) It's tricky to plot graphs with **very small** and **very large** numbers (e.g. both 0.1 and 1000) on the **same axis**.
2) You can make it easier by converting values to their **logarithms** and plotting them on a **logarithmic scale** (e.g. a $\log_{10}$ scale).
3) On a **$\log_{10}$ scale**, each value is **ten times larger** than the value before. This means the numbers 1, 2, 3 and 4 on a **$\log_{10}$ scale** represent 10, 100, 1000 and 10 000 on a **linear (normal) scale**.

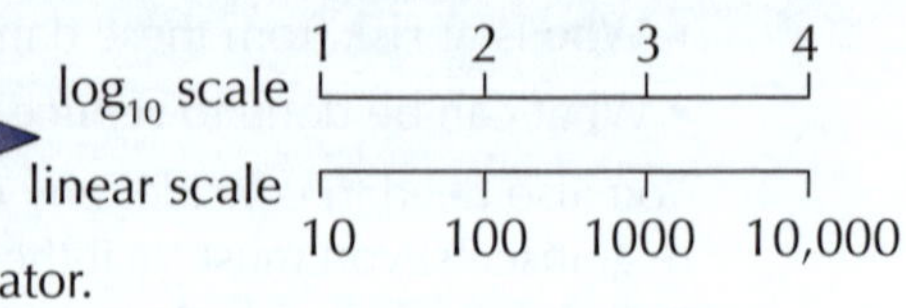

4) To calculate logarithms, you need to use the **log button** on your calculator. On most calculators 'log' will stand for $\log_{10}$, but different calculators work differently, so make sure you know how to use yours.

Averages and the **Range** Can be Used to **Summarise** Your **Data**

1) When you've done **repeats** of an experiment you should always calculate a **mean** (a type of average). To do this **add together** all the data values and **divide** by the **total** number of values in the sample.

Test tube	Mass (g)			Mean (g)	Range (g)
	Repeat 1	Repeat 2	Repeat 3		
A	28	37	32	(28 + 37 + 32) ÷ 3 = 32.3	37 – 28 = 9
B	47	51	60	(47 + 51 + 60) ÷ 3 = 52.7	60 – 47 = 13

2) You might also need to calculate the **range** (how **spread out** the data is). To do this find the **largest** data value and **subtract** the **smallest** data value from it.
3) **Standard deviation** can be more useful than the **range** because it tells you how **values** are spread about the **mean** rather than just the **total spread** of data. A **small standard deviation** means the repeated results are all **similar** and **close** to the mean, i.e. they are **precise**.

Like the mean, the **median** and **mode** are both types of average.

- To calculate the **median**, put all your data in **numerical order**. The median is the **middle value** in this list. If you have an **even number** of values, the median is **halfway** between the middle two values.
- To calculate the **mode**, count **how many times** each value comes up. The mode is the number that appears **most often**. A set of data might not have a mode — or it might have more than one.

Processing and Presenting Data

PRACTICAL SKILLS

Watch Out For **Significant Figures...**

1) The **first significant figure** of a number is the **first digit** that **isn't a zero**. The second, third and fourth significant figures follow on immediately after the first (even if they're zeros).
2) When you're processing your data you may well want to round any **really long numbers** to a certain number of **significant figures**. E.g. **0.6878976** rounds to **0.69** to **2 s.f.**.
3) When you're doing **calculations** using measurements given to a certain number of significant figures, you should give your **answer** to the **lowest number** of significant figures that was used in the calculation. For example:

$1.2 \div 1.85 = 0.648648648...$ **= 0.65**

2 s.f. (1.2) — 3 s.f. (1.85) — Answer should be rounded to 2 s.f. — Round the last digit up to 5.

When rounding a number, if the next digit after the last significant figure you're using is less than five, you should round it down and if it's 5 or more you should round it up.

4) This is because the **fewer digits** a measurement has, the less **accurate** it is. Your answer can only be as accurate as the **least accurate measurement** in the calculation.

...and **Standard Form**

1) When you're processing data you might also want to change **very big** or **very small numbers** that have **lots of zeros** into something more manageable — this is called **standard form**.

E.g. 1 000 000 can be written 1×10^6 and 0.017 can be written 1.7×10^{-2}.

A rabbit playing the piano. Definitely not standard form.

2) To do this you just need to **move the decimal point** left or right. The number of places the decimal point moves is then represented by a **power of 10** — this is positive for big numbers, and negative for numbers smaller than one. For example:

$16\,500 = 1.65 \times 10^4$ The decimal point has moved **four places** to the **left**, so the power of 10 is **+4**.

$0.000362 = 3.62 \times 10^{-4}$ The decimal point has moved **four places** to the **right**, so the power of 10 is **–4**.

You Need to Understand How and When **Statistical Tests** are Used to **Analyse Data**

Examples:

1) The **Student's t-test**. You can use the Student's t-test when you have two sets of **data** that you want to **compare**. It tests whether there is a **significant difference** in the **means** of the two data sets. The value obtained is compared to a **critical value**, which helps you decide how likely it is that the results or 'differences in the means' were **due to chance**. If the value obtained from the t-test is **greater than** the critical value at a **probability** (**P value**) of **5% or less** (≤ 0.05), then you can be 95% confident that the difference is significant and not due to chance. This is called a **95% confidence limit** — which is good enough for most biologists to **reject** the **null hypothesis**. A null hypothesis is a special type of hypothesis used with statistical tests. It states that there's no significant difference between the things you're measuring.
2) The **Chi-squared test** (see pages 64-65). You can use the Chi-squared test when you have **categorical** (grouped) **data** and you want to compare whether your **observed results** are **statistically different** from your **expected results**. You compare your result to a **critical value** — if it's **larger** than the critical value at **P = 0.05**, you can be **95% certain** the difference is significant.
3) A correlation coefficient, e.g. the **Spearman's rank correlation coefficient**. This test allows you to work out the **degree** to which **two** sets of **data** are **correlated** (see page 119 for more on correlation). It is given as a value between 1 and –1. A value of 1 indicates a **strong positive correlation**, 0 means there is **no correlation** and –1 is a **strong negative correlation**. You can then compare your result to a critical value to find out whether or not the correlation is significant.

You can be more confident in your **conclusions** (see page 119), if they're based on results that have been analysed using a statistical test.

You need to be familiar with the symbols < (less than), > (more than), << (much less than) and >> (much greater than).

PRACTICAL SKILLS

Processing and Presenting Data

Use a Suitable **Graph** or **Chart** to **Present** Your **Data**

Graphs and charts are a great way of **presenting data** — they can make results much **easier to interpret**.

1) When you have **qualitative** data (non-numerical data, e.g. blood group) or **discrete** data (numerical data that can only take certain values in a range, e.g. shoe size) you can use **bar charts** or **pie charts**.
2) When you have **continuous** data (data that can take any value in a range, e.g. height or weight) you can use **histograms** or **line graphs.**
3) When you want to show how **two variables** are **related** (or **correlated**, see next page) you can use a **scatter graph**.

Scatter graph:

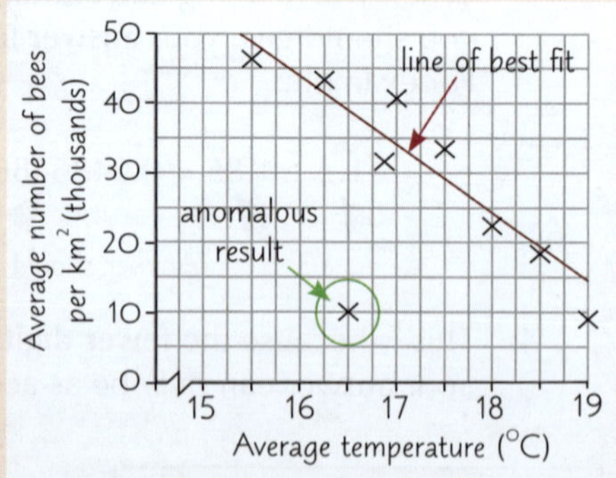

Whatever type of graph you use, you should make sure that:

- The **dependent variable** goes on the **y-axis** (the vertical axis) and the **independent** on the **x-axis** (the horizontal axis).
- You always **label** the **axes**, include the quantity and **units**, and choose a **sensible scale**.
- The graph covers **at least half** of the **graph paper**.

If you need to draw a **line** (or curve) **of best fit** on a **scatter graph**, draw the line through or as near to as many points as possible, **ignoring** any **anomalous** results.

Find the **Rate** By Finding the **Gradient**

Rate is a **measure** of how much something is **changing over time**. Calculating a rate can be useful when analysing your data, e.g. you might want to the find the **rate of a reaction**. Rates are easy to work out from a graph:

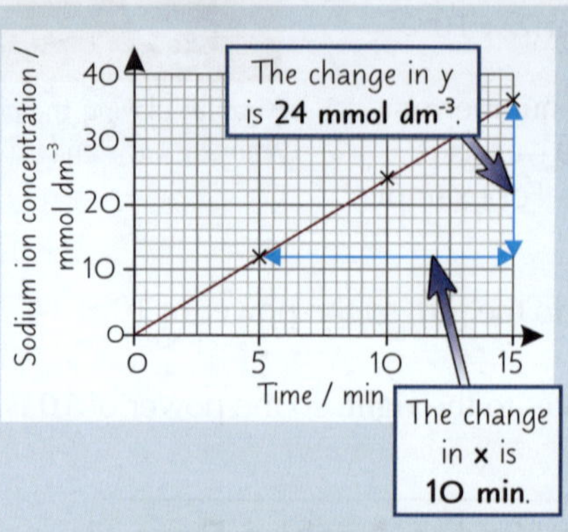

For a **linear** graph you can calculate the **rate** by finding the **gradient of the line**:

$$\text{Gradient} = \frac{\text{Change in Y}}{\text{Change in X}}$$

So in this **example**:

$$\text{rate} = \frac{24 \text{ mmol dm}^{-3}}{10 \text{ minutes}} = \mathbf{2.4 \text{ mmol dm}^{-3} \text{ min}^{-1}}$$

The **equation** of a **straight line** can always be written in the form **$y = mx + c$**, where **m** is the **gradient** and **c** is the **y-intercept** (this is the **value of y** when the line crosses the **y-axis**). In this example, the equation of the line is **$y = 2.4x + 0$** (or just **$y = 2.4x$**). Knowing the equation of the line allows you to estimate results not plotted on the graph. E.g. in this case, when x (the time) is **20 min**, y (the sodium ion concentration) will be $2.4x = 2.4 \times 20 =$ **48 mmol dm^{-3} min^{-1}**.

For a **curved** (non-linear) graph you can find the **rate** by drawing a **tangent**:

1) Position a ruler on the graph at the **point** where you want to know the **rate**.
2) **Angle** the **ruler** so there is **equal space** between the **ruler** and the **curve** on **either** side of the point.
3) **Draw** a **line** along the ruler to make the tangent. Extend the line right across the graph — it'll help to make your gradient calculation easier as you'll have more points to choose from.
4) **Calculate** the **gradient** of the **tangent** to find the **rate**.

Gradient = 55 m² ÷ 4.4 years = **12.5 m² year^{-1}**

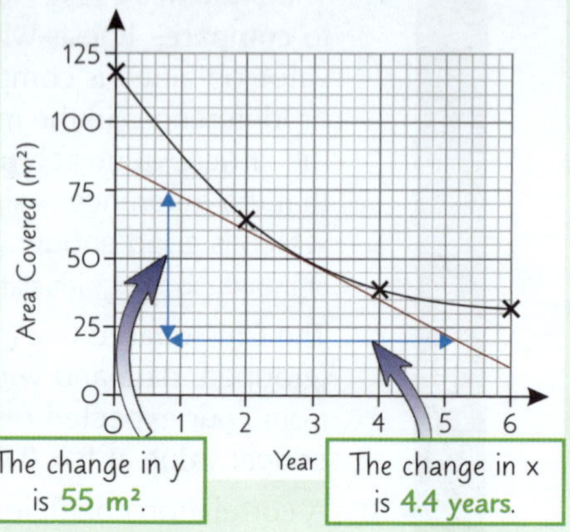

When calculating a rate (or anything else for that matter) you might have to **convert** between **units**, e.g. seconds and minutes. Make sure you can convert between common units of time, length and volume.

seconds → ÷ 60 → minutes; minutes → × 60 → seconds

Significant figures — a result of far too many cream cakes...

Lots of maths to get your head around on these three pages, but stay calm and take your time with it all. You'll be fine.

Drawing Conclusions and Evaluating

PRACTICAL SKILLS

There's no point in getting all those lovely results and just leaving it at that. You need to draw some conclusions...

You Need to be Able to **Draw Conclusions** From **Data**

1) Conclusions need to be **valid**. A conclusion can only be considered as valid if it uses valid data (see page 114).
2) You can often draw conclusions by looking at the relationship (**correlation**) between two variables:

Positive correlation
As one variable increases the other increases.

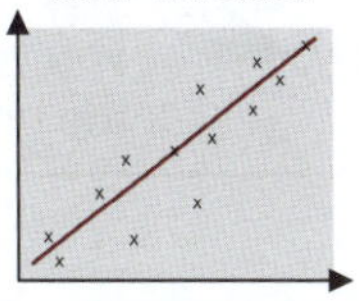

Negative correlation
As one variable increases the other decreases.

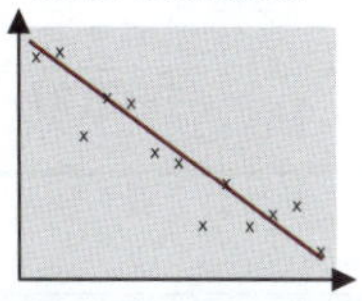

No correlation
There is no relationship between the variables.

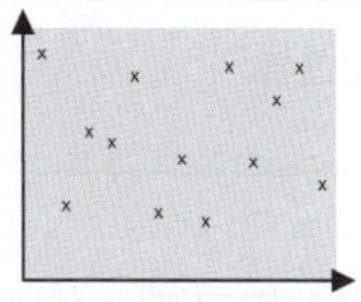

There is no correlation between the colour of your tights and the proportion of your life you spend upside down.

3) You have to be very **careful** when **drawing conclusions** from data like this because a **correlation** between two variables **doesn't** always mean that a **change** in one variable **causes** a **change** in the other (the correlation could be due to **chance** or there could be a **third variable** having an effect).
4) If there's a relationship between two variables and a change in one variable **does** cause a change in the other it's called a **causal relationship**.
5) It can be **concluded** that a **correlation** is a **causal relationship** if every other variable that could possibly affect the result is **controlled**.
6) When you're making a conclusion you **can't** make broad **generalisations** from data — you have to be very **specific**. You can only **conclude** what the results show and **no more**.

In reality this is very hard to do — correlations are generally accepted to be causal relationships if lots of studies have found the same thing, and scientists have figured out exactly how one factor causes the other.

Example

The graph shows the results from an investigation into the effect of concentration of plant growth factor X on the height of Plant Species A. The only **conclusion** you can draw is that as the **concentration** of **growth factor X increases**, the **height** of **Plant Species A increases**. You **can't** conclude that this is true for any other plant growth factor or any other plant species — the **results** could be **completely different**.

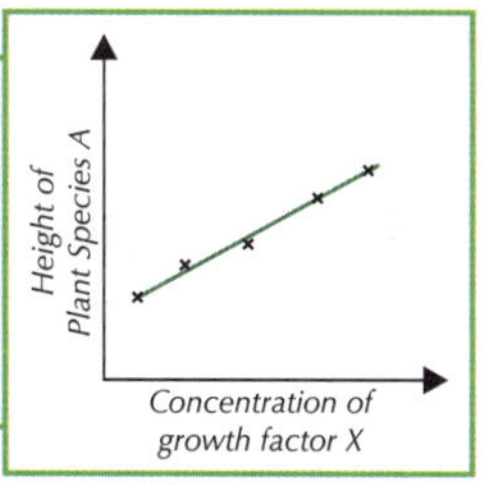

Uncertainty is the Amount of **Error** Your **Measurements** Might Have

1) The results you get from an experiment won't be completely perfect — there'll always be a **degree of uncertainty** in your measurements due to limits in the **sensitivity** of the apparatus you're using.
2) For example, an electronic mass balance might measure to the **nearest 0.01 g**, but the real mass could be up to **0.005 g smaller or larger**. It has an **uncertainty value** of **± 0.005 g**.
3) A ± sign tells you the **range** in which the **true value** lies (usually to within a **95% confidence level**). The range is called the **margin of error**.

You Can **Calculate** The **Percentage Error** of Your **Measurements**

If you know the **uncertainty value** of your measurements, you can calculate the **percentage error** using this formula:

$$\text{percentage error} = \frac{\text{uncertainty}}{\text{reading}} \times 100$$

Example

50 cm^3 of HCl is measured with an uncertainty value of ± 0.05 cm^3.

percentage error = $\frac{0.05}{50} \times 100 = \mathbf{0.1\%}$

PRACTICAL SKILLS

Drawing Conclusions and Evaluating

You Can **Minimise** the **Errors** in Your **Measurements**

1) One obvious way to **reduce errors** in your measurements is to buy the most **sensitive equipment** available. In real life there's not much you can do about this one — you're stuck with whatever your school or college has got. But there are other ways to **lower the uncertainty** in experiments.
2) For example, you can plan your experiment so you **measure** a **greater amount** of something:

If you use a **500 cm³** cylinder that goes up in **5 cm³** increments, each reading has an uncertainty of **± 2.5 cm³**.

So using a 500 cm³ cylinder to measure **100 cm³** of liquid will give you a percentage error of:

$$\frac{2.5}{100} \times 100 = \mathbf{2.5\%}$$

But if you measure **200 cm³** in the same cylinder, the percentage error is:

$$\frac{2.5}{200} \times 100 = \mathbf{1.25\%}$$

Hey presto — you've just **halved** the uncertainty.

You Also Need to Be Able to **Evaluate Methods** and **Results**

1) Here are some things to **think about** when evaluating experimental results:
   - **Repeatability**: Did you take enough repeat readings of the measurements? Would you do more repeats if you were to do the experiment again? Do you think you'd get similar data if you did the experiment again?
   - **Reproducibility**: Have you compared your results with other people's results? Were your results similar? Could other scientists gain data showing the same relationships that are shown in your data?
   - **Validity**: Does your data answer the question you set out to investigate? Were all the variables controlled?
2) Make sure you **evaluate** your **method** too. Is there anything you could have done to make your results more **precise** or **accurate**? Were there any **limitations** in your method, e.g. should you have taken measurements more **frequently**? Were there any **sources** of **error** in your experiment? Could you have used more sensitive **apparatus** or **equipment**? Think about how you could **refine** and **improve** your experiment if you did it again.
3) Once you've thought about these points you can decide how much **confidence** you have in your **conclusion**. For example, if your results are **repeatable**, **reproducible** and **valid** and they back up your conclusion then you can have a **high degree** of **confidence** in your conclusion.

You can apply all these questions to any results or methods you're given to evaluate in the exams too.

Solving Problems in a **Practical Context**

In the exams, you'll get plenty of questions set in a 'practical context'. As well as answering questions about the methods used or the conclusions drawn, you'll need to be able to **apply** your **scientific knowledge** to **solve problems** set in these contexts. For example:

Q1 An experiment was carried out to investigate the role of IAA in shoot growth. The experimental set up is shown in the diagram on the right.

Four shoots were then placed in the dark (experiment 1) and the other four shoots were exposed to a light source directed from the right (experiment 2). After two days, the amount of growth (in mm) and direction of growth was recorded. The results are shown in the table.

a) Explain the results seen for shoot C. [3 marks]

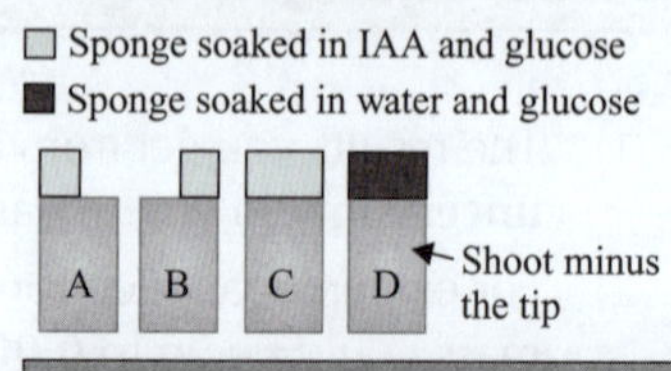

	Growth / mm			
	Shoot A	Shoot B	Shoot C	Shoot D
Experiment 1 (dark)	6, right	6, left	6, straight	1, straight
Experiment 2 (light)	8, right	8, right	8, right	3, straight

You should remember from page 30 that **IAA** stimulates **cell elongation** in a shoot. In experiment 1, equal amounts of IAA diffuse down **both sides** of shoot C, making the cells elongate at the **same rate**, so the shoot grows **straight up**. In experiment 2, IAA moved to the **shaded** (left-hand side) of shoot C, so the shoot grew to the **right** — towards the **light**.

Correlation Street — my favourite programme...

Don't ever, ever assume that correlation means cause. There, I've told you again. No excuses now. A good evaluation is a sign that you really understand what makes a good experiment, so make sure your evaluation-writing-skills are top notch.

How to Do Well in Your Exams

The reason for learning all the lovely facts and diagrams in this book is so that you can ace your exams and get yourself an A-level in Biology. So, now it's a good idea to find out exactly what you'll be in for exam-wise...

Make Sure You Know the **Structure** of Your **Exams**

It seems obvious, but if you know exactly what will be **covered** in each of the exams, how much **time** you'll have to do them and how they'll be **structured**, you can be better prepared. So let's take a look at the ins and outs of all the exams you'll be facing for **A-level Biology**...

Paper	Total marks	Time	Topics assessed
1	91	2 hours	1, 2, 3, 4
2	91	2 hours	5, 6, 7, 8
3	78	2 hours	1 to 8

All three A-level papers also test you on Practical Skills — see pages 114-120 for more.

This book covers **Topics 5 to 8** — the material from **Year 2** of your course. You'll have covered **Topics 1 to 4** in **Year 1** of your course, so make sure you **revise your Year 1 notes** for the **Paper 1** and **Paper 3 exams**.

1) **Papers 1** and **2** are mainly a mixture of **short** and **long answer questions**. Some of these questions will test you on the **facts** you need to know, some will test whether you can **apply your knowledge** to unfamiliar contexts and some will test your knowledge of **practical skills**. There'll also be a few **calculation questions**.
2) **Paper 1** also contains **15 marks' worth** of **extended response questions**. These are questions that require you to write a **longer answer** with a **logical structure**. E.g. you could be asked to describe the **steps** in a particular **process**. These questions could involve an **extended calculation** too.
3) **Paper 2** also contains a **15 mark comprehension question**. You'll be given a **passage of information** to **read** and will then need to **answer** the question parts that follow using **both** the **information** you've been given and your **own scientific knowledge**.
4) **Paper 3** is split into two sections. **Section A** has lots of questions on **practical techniques** and **skills**, with 15 marks being awarded for questions that ask you for a **critical analysis** of **experimental data**. For example, you could be given some data (e.g. in a graph or table) and asked to **draw conclusions** from it or you could be given a conclusion and asked to **evaluate** how well the data **supports the conclusion**. As for Papers 1 and 2, there'll also be fact recall questions, questions that test whether you can apply your knowledge, and calculation questions.
5) **Section B** of **Paper 3** consists of a **25 mark synoptic essay question**...

Synoptic means you will need to draw together your knowledge of different areas of Biology in relation to a theme.

You Need to be Able to **Write** a **Good Essay**

You'll be given a **choice** of **two** essay titles in Section B of Paper 3 and asked to write about **one** of them. The titles are designed to get you to write about a **range of material** from **both years** of your A-level course. Writing an essay might seem like a **daunting task**, but **don't panic**. Here are some tips for getting **top marks**:

1) Before you start your essay, it's a good idea to quickly scribble down a **rough plan** — this should help you to present your ideas in a **clear, logical way**. It should also **stop you** from **repeating yourself** or **missing out** any **important bits**. You should aim to write about **at least five different topic areas**.
2) You'll need to clearly show how **all** the information you include is **relevant** to the **essay title** — don't just write down everything you know about a topic.
3) The information you include must be **detailed**, **scientifically correct** and of **A-level standard**. 'Plants are green and have leaves' just ain't gonna cut it I'm afraid...
4) You must use appropriate **scientific terminology**.
5) Your essay should be **well-written** and **clearly explained**.
6) To get the **very highest marks**, your answer should show evidence of **wider reading** (i.e. it should include things that aren't explicitly on the specification, but are still of a high standard and relevant to the question).

Making a quick plan of the topics you'll cover might help you decide which title you can write a better essay for.

You'll get 2 hours in total for this paper and should aim to leave yourself **about 50 minutes** to plan and write your essay. This should be enough time to write **about 3 sides of A4**.

How to Do Well in Your Exams

Command Words Tell You What You Need to do in a Question

Command words are just the bits of a question that tell you **what to do**. You'll find answering exam questions much easier if you understand exactly what they mean, so here's a brief summary table of the **most common** ones:

Command word:	What to do:
Give / Name / State	Give a brief one or two word answer, or a short sentence.
Describe	Write about what something's like, e.g. describe the structure of fish gills.
Explain	Give reasons for something.
Suggest	Use your scientific knowledge to work out what the answer might be.
Compare	Give the similarities and differences between two things.
Contrast	Give the differences between two things.
Calculate	Work out the solution to a mathematical problem.
Evaluate	Give the arguments both for and against an issue, or the advantages and disadvantages of something. You also need to give an overall judgement.

Even though you're taking an A-level in Biology, there will be some maths to do in these papers that's set in a biological context.

Some questions will also ask you to answer '**using the information/data provided**' (e.g. a graph, table or passage of text) or '**with reference to figure X**' — if so, you must **refer to** the information, data or figure you've been given or you won't get the marks. Some questions may also ask you to answer '**using your calculation**' — it's the same here, you need to use your **answer** to a particular **calculation**, otherwise you won't get the marks.

Not all of the questions will have command words — instead they may just ask a which / what / how type of question.

Time Management is Important

1) For **Papers 1** and **2**, you get **just over a minute per mark**. So if you get stuck on a short question, it's sometimes worth moving on to another one and then coming back to it if you have time. Bear in mind that you might want to spend a **bit longer** than a minute per mark on the **extended response** and **comprehension questions**.
2) For **Paper 3**, it's a **similar story** — you'll want to **spend longer per mark** on the **essay question** than on the shorter questions, so make sure you **leave enough time** for this at the end.

Warm-Up Questions

Q1 Which two A-level papers test you on material from Topics 1 to 4?

Q2 Which A-level papers test you on Practical Skills?

Q3 What's the difference between the command words 'describe' and 'explain'?

PRACTICE QUESTIONS

Exam Question

Q1 Write an essay about the importance of proteins to living organisms. [25 marks]

You might think you need your head examined for picking A-level Biology...

...because there's a lot to learn and three big exams to do. But let me just stop you right there... Instead of worrying, just work through this book, including having a go at all of the questions and you'll be well and truly prepped for the exams. Then re-read these pages to make sure you know what's coming. After that, all there is to say is... good luck.

Synoptic Practice

If you've ticked off everything else there is to do in this book then you've come to the right place. To answer synoptic questions you'll need to pull together your knowledge of different topics, which can be pretty tough going. Luckily, these pages are full of synoptic questions for you to have a go at...

1 A pyrogen is any substance that causes a fever, which is an increase in the body temperature above 38 °C. Substances produced by some bacteria can act as pyrogens in an infected host.

1.1 During a fever, the body shivers, which is the involuntary, rapid contraction and relaxation of muscles. Explain how shivering may affect the rate of glycolysis in the body.

(2 marks)

1.2 Shivering increases the oxygen requirement of the muscle cells. Explain how this increase in oxygen demand could cause heart rate to increase.

(5 marks)

Scientists investigated the effects of several substances in rodents in order to determine if they could act as pyrogens in humans. They tested each substance (**A-D**) on 20 different rodents and recorded the average increase in body temperature they caused. Their results are shown in **Table 1**.

Table 1

Substance	Average increase in body temperature / °C	Range / °C	Standard deviation / °C
A	0.3	0.5	± 0.2
B	1.1	0.6	± 0.3
C	1.2	1.1	± 0.6
D	0.9	0.7	± 0.3

1.3 The range and standard deviation are both values that indicate how the results are dispersed. Explain which measure is better for judging the precision of the results shown in **Table 1**.

(2 marks)

1.4 The greater the increase in body temperature recorded, the more dangerous the scientists considered the substance to be. Using data from **Table 1**, calculate the percentage difference in average increase in body temperature between the most and the least dangerous substance.

(1 mark)

2 Neuromuscular blockers are a group of muscle relaxant drugs often used to ensure patients do not move during surgery.

2.1 One such drug prevents acetylcholine from binding to nicotinic cholinergic receptors. Explain how this neuromuscular blocker would prevent muscle contraction.

(5 marks)

2.2 Another type of neuromuscular blocker works by binding to nicotinic cholinergic receptors and generating an action potential. The drug remains bound to the receptors for a long time, resulting in extended depolarisation of the postsynaptic membrane.
Using your knowledge of how an action potential works, explain how this drug prevents a person from moving.

(2 marks)

2.3 Other than binding to nicotinic cholinergic receptors, suggest **one** way in which a drug could act at a neuromuscular junction to prevent a person from moving.

(1 mark)

Synoptic Practice

Scientists investigated how the dose of a neuromuscular blocker given to patients affected the force of muscle contraction. In their investigation, the scientists measured the force of contraction of one muscle in the body before the patient had been given any of the drug and again once they had received a dose of the drug. 80 patients were included in the investigation. The results are shown in **Figure 1**.

Figure 1

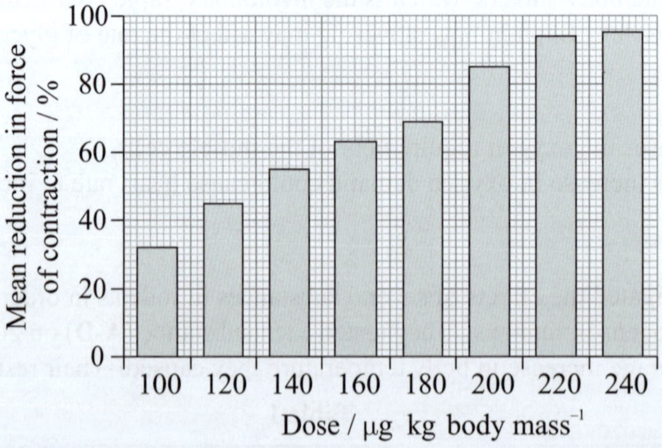

2.4 For the drug to be classed as suitable for use in surgery, the scientists recommended that the patient should have no more than 15% of their usual force of contraction remaining. Using this information and **Figure 1**, calculate the minimum amount of the drug in mg that should be given to a patient with a body mass of 75 kg.

(2 marks)

2.5 A student wanted to determine if the difference in the results for 200 µg kg body mass^{-1} and 220 µg kg body mass^{-1} were statistically significant.
Suggest which statistical test they could use to determine this. Explain your choice.

(2 marks)

2.6 The kidneys play a role in the excretion of drugs from the body. Some neuromuscular blocker molecules entering the kidneys from the renal arteries can pass into the Bowman's capsules to be excreted. However, the number of functioning glomeruli a person has tends to decline with age, meaning doctors need to consider a patient's age before administering neuromuscular blockers.
Using the information provided, suggest why a neuromuscular blocker may have a prolonged effect in an elderly patient.

(2 marks)

2.7 After surgery, patients can be given anticholinesterases to help overcome the effects of the neuromuscular blockers. Anticholinesterases have an inhibitory effect on the enzyme, acetylcholinesterase, which normally breaks down acetylcholine in the synaptic cleft.
Suggest how anticholinesterases work to overcome the effects of the neuromuscular blockers.

(2 marks)

3 Familial hypercholesterolaemia (FH) is an inherited medical condition that is characterised by a high blood cholesterol level. Normally, low density lipoproteins (LDLs) carry cholesterol molecules through the blood and into body cells, by binding to LDL receptors present on cells. FH can be caused by many mutations that affect the LDL receptor gene and result in defective LDL receptors, causing cholesterol to remain in the blood.

3.1 FH can be described as an autosomal dominant disease.
Describe what is meant by the term 'autosomal dominant'.

(2 marks)

Synoptic Practice

3.2 There are several mutations that can lead to the LDL receptor being defective. An individual was screened for mutations in their LDL receptor gene that would put them at risk of FH. They were found to have a mutation that resulted in a change from the triplet code CCG to CTG. No other mutations were detected. Explain whether this mutation would cause a frameshift.

(2 marks)

Figure 2 shows the incidence of FH in some members of a family. Horizontal lines between two individuals link mating parents and vertical lines link parents and offspring.

Figure 2

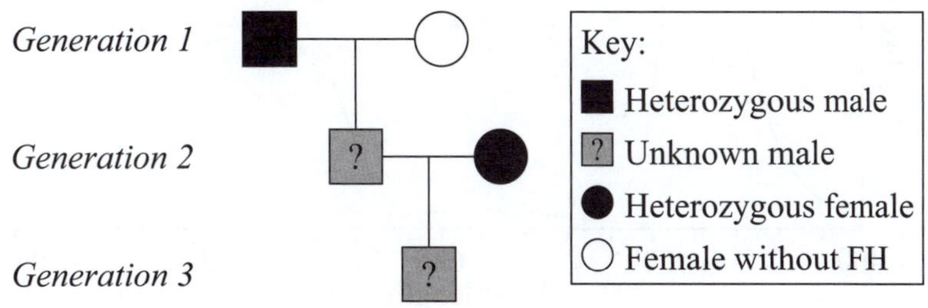

3.3 One student concluded that the Generation 3 male has a 50% chance of inheriting FH. Evaluate this conclusion.

(3 marks)

It is estimated that around 1 in every 1 000 000 people in the UK is homozygous for the FH allele.

3.4 A city in the UK has an estimated population size of 9.62×10^6 people. Calculate an estimate of how many people in this UK city are homozygous for the FH allele.

(1 mark)

3.5 Using the Hardy-Weinberg equations, calculate the frequency of the heterozygous genotype in the UK. Show your working.

(2 marks)

People with FH can be given drugs, called statins, to reduce their blood cholesterol level. However, there are some side-effects associated with statins, such as an increased risk of developing Type II diabetes.

3.6 Statins can increase the expression a protein complex called NLRP3 inflammasome. This protein complex can lead to the inflammation of β cells in the pancreatic islets of Langerhans. Suggest an explanation for how this could lead to the development of Type II diabetes.

(4 marks)

Scientists hope that gene therapy could eventually be used as an alternative treatment for FH.

3.7 Explain how gene therapy could be used to increase the uptake of cholesterol from the blood of a person with FH.

(4 marks)

3.8 Suggest **one** possible complication of the use of gene therapy as a treatment for FH.

(1 mark)

4 Light is an important abiotic factor for optimum crop plant growth.

4.1 Over several years, large areas of a farmer's crop fields have become more shaded. Rabbits are a pest species that feed on the farmer's crop plants. Explain how the increase in shade may have changed the carrying capacity of the rabbits living in the fields.

(3 marks)

Synoptic Practice

Plant leaves contain photosynthetic pigments, which can absorb different wavelengths of light. **Figure 3** shows the absorption spectra for two photosynthetic pigments found in plant leaves, chlorophyll a and chlorophyll b.

Figure 3

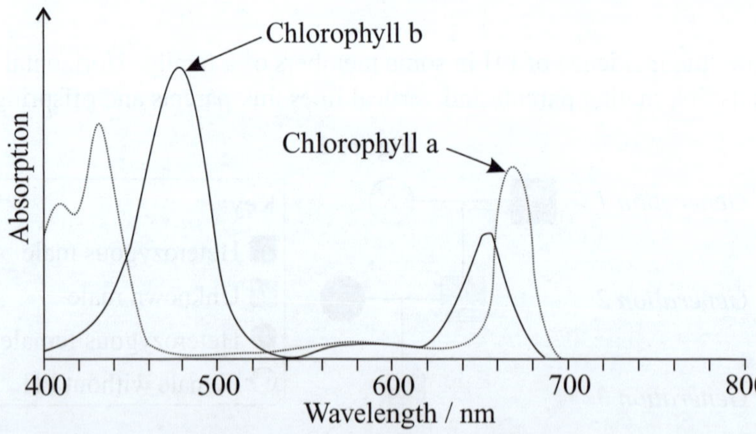

Far-red light is light with a wavelength of around 700-800 nm.

4.2 Using **Figure 3**, explain why exposure to only far-red light could limit plant growth.

(3 marks)

4.3 Far-red light is reflected off the leaves of plants. Plants can detect an increase in the amount of far-red light, which indicates that their environment may have become more shaded. Far-red light is detected by a photoreceptor called phytochrome. When phytochrome is exposed to far-red light, it is converted from an active form (Pfr) into an inactive form (Pr). Pfr inhibits the activity of a protein called TTA1, which is involved in the synthesis of indoleacetic acid.
Suggest and explain how the detection of far-red light could help a plant to maximise growth as its environment becomes more shaded.

(3 marks)

4.4 When different individuals of the same plant species have been grown in either shaded or brightly lit environments, they have been observed to develop different phenotypes.
Explain how transcription factors can lead to differences in phenotypes between two organisms that have the same genome.

(3 marks)

5 Read the following passage.

Huntington's disease is an inherited disorder for which there is currently no cure. The gene responsible for the disease, *HTT*, contains a sequence of CAG triplet repeats. An autosomal dominant mutation in the *HTT* gene can lead to many additional CAG triplet repeats being added to this DNA sequence. The huntingtin protein produced as a result of this mutation can lead to the degradation of neurones in the brain, which can lead to a reduction in the secretion of inhibitory neurotransmitters by these neurones.

Symptoms of Huntington's disease often do not develop until adulthood, and many affected individuals have children prior to a diagnosis. Those who are at risk from Huntington's disease, such as those with parents who have the disease, may decide to have a genetic test to determine if they have the mutation.

Synoptic Practice

During the genetic test, a blood sample is taken from a person and the DNA isolated. PCR can then be used to create many copies of the *HTT* gene present in the DNA. Electrophoresis can then be carried out on the products of PCR, and the results used to help determine whether the person is likely to develop Huntington's disease.

A new drug has been developed that shows promise of slowing down the progression of Huntington's disease. The drug functions by relying on the basic principles of RNA interference.

Using information from the passage and your own knowledge, answer the following questions.

5.1 Huntington's disease is caused by an autosomal dominant mutation (line 3).
A man is heterozygous for the *HTT* gene and his wife is homozygous recessive.
Give the probability that their child will inherit Huntington's disease.

(1 mark)

5.2 The mutation in the *HTT* gene can lead to many additional CAG triplet repeats being added to the DNA sequence (lines 3-4).
Suggest why this mutation leads to the formation of a defective huntingtin protein.

(2 marks)

5.3 Huntington's disease can lead to a reduction in the secretion of inhibitory neurotransmitters in the brain (lines 5-6).
Explain how inhibitory neurotransmitters can prevent the transmission of a nerve impulse.

(2 marks)

5.4 People may decide to have a genetic test to determine if they have the mutation responsible for Huntington's disease (lines 9-10).
Give **one** reason for and **one** reason against having the genetic test for Huntington's disease.

(2 marks)

5.5 PCR can be used to create many copies of the *HTT* gene (line 12).
Explain why DNA needs to be temporarily heated to 95 °C during the PCR procedure.

(2 marks)

5.6 Electrophoresis can be carried out on the products of PCR in the genetic test for Huntington's disease (line 13).
Suggest how electrophoresis could be used to determine if a person is at risk of developing Huntington's disease.

(2 marks)

5.7 The new drug developed for Huntington's disease functions by relying on the basic principles of RNA interference (lines 16-17).
Explain how RNA interference could prevent the development of Huntington's disease.

(4 marks)

6 Write an essay on the topic below.

The importance of enzymes in living organisms.

(25 marks)

Answers

Topic 5A — Photosynthesis and Respiration

Page 3 — Photosynthesis, Respiration and ATP

1 Any six points from: e.g. in the cell, ATP is synthesised from ADP and inorganic phosphate/P_i ***[1 mark]*** using energy from an energy-releasing reaction, e.g. respiration ***[1 mark]***. The energy is stored as chemical energy in the phosphate bond ***[1 mark]***. ATP synthase catalyses this reaction ***[1 mark]***. ATP then diffuses to the part of the cell that needs energy ***[1 mark]***. Here, it's broken down back into ADP and inorganic phosphate/P_i ***[1 mark]***, which is catalysed by ATP hydrolase ***[1 mark]***. Chemical energy is released from the phosphate bond and used by the cell ***[1 mark]***.
Make sure you don't get the two enzymes confused — ATP **syn**thase **syn**thesises ATP, and ATP hydrolase breaks it down.

Page 7 — Photosynthesis

1 a) Photosystem II ***[1 mark]***.
b) Photolysis/light energy ***[1 mark]*** splits water into two hydrogen ions and oxygen ***[1 mark]***. The electrons from the water replace the electrons lost from chlorophyll ***[1 mark]***.
The question asks you to explain the purpose of photolysis, so make sure you include why the water is split up — to replace the electrons lost from chlorophyll.
c) Excited electrons are transferred to reactant D/NADP from photosystem I/object C ***[1 mark]*** along with a proton/H^+ ion from the stroma ***[1 mark]***.

2 a) Any five points from: e.g. ribulose bisphosphate/RuBP and carbon dioxide/CO_2 join together to form an unstable 6-carbon compound ***[1 mark]***. This reaction is catalysed by the enzyme rubisco ***[1 mark]***. The compound breaks down into two molecules of a 3-carbon compound called glycerate 3-phosphate/GP ***[1 mark]***. Two molecules of glycerate 3-phosphate are then converted into two molecules of triose phosphate/TP ***[1 mark]***. The energy for this reaction comes from ATP ***[1 mark]*** and the H^+ ions come from reduced NADP ***[1 mark]***.
b) Ribulose bisphosphate is regenerated from triose phosphate/TP molecules ***[1 mark]***. ATP provides the energy to do this ***[1 mark]***.
This question is only worth two marks so only the main facts are needed, without the detail of the number of molecules.
c) No glycerate 3-phosphate/GP would be produced ***[1 mark]***, so no triose phosphate/TP would be produced ***[1 mark]***. This means there would be no glucose produced ***[1 mark]***.

Page 9 — Limiting Factors in Photosynthesis

1 a) Any two points from: e.g. by burning propane to increase air CO_2 concentration ***[1 mark]***. / By adding heaters to increase temperature ***[1 mark]***. / By adding coolers to decrease temperature ***[1 mark]***. / By adding lamps to provide light at night ***[1 mark]***.
b) Potatoes ***[1 mark]*** because the yield showed the smallest percentage increase of 25% ($850 - 680 = 170$, $170 \div 680 \times 100 = 25\%$) ***[1 mark]***.

Page 11 — Photosynthesis Experiments

1 a) Dehydrogenase enzymes catalyse the reaction that produces reduced NADP ***[1 mark]***.
b) Redox indicator dyes take the place of NADP as an electron acceptor ***[1 mark]***. This means that dehydrogenase activity reduces the dye instead of NADP ***[1 mark]***. The reduction is coupled with a colour change, which can be easily observed ***[1 mark]***.

Page 15 — Aerobic Respiration

1 a) Reduced NAD ***[1 mark]***
b) The regenerated NAD is needed for glycolysis to continue ***[1 mark]*** and ATP to be produced under anaerobic conditions, providing the energy to keep running ***[1 mark]***.

2 a) The transfer of electrons down the electron transport chain stops ***[1 mark]***. So there's no energy released to phosphorylate ADP/produce ATP ***[1 mark]***.
b) The Krebs cycle stops ***[1 mark]*** because there's no oxidised NAD/FAD coming from the electron transport chain ***[1 mark]***.
Remember that when the electron transport chain is inhibited, the reactions that depend on the products of the chain are also affected.

3 Any six points from: e.g. glucose is phosphorylated using a molecule of ATP ***[1 mark]***. This creates one molecule of glucose phosphate ***[1 mark]*** and one molecule of ADP ***[1 mark]***. ATP is used to add another phosphate to glucose phosphate ***[1 mark]***, forming hexose bisphosphate ***[1 mark]***, which is then split into two molecules of triose phosphate ***[1 mark]***. Triose phosphate is oxidised/loses hydrogen to form two molecules of pyruvate ***[1 mark]***. NAD collects the hydrogen ions, forming two molecules of reduced NAD ***[1 mark]***.

Page 17 — Respiration Experiments

1 a) To stop oxygen getting into the solution, which forces the yeast to respire anaerobically ***[1 mark]***.
b) Both ethanol and CO_2 are products of anaerobic respiration ***[1 mark]***. Measuring how fast CO_2 is produced would indicate how fast ethanol is being produced ***[1 mark]***.
c) Any two from: e.g. the temperature the investigation is being carried out at — could be controlled by putting the test tubes in a water bath at a set temperature ***[1 mark]***. / The mass of yeast used — could be controlled by weighing out a set amount of yeast to use in each test tube ***[1 mark]***. / The volume/concentration of the glucose solution used — could be controlled by measuring out a known volume of glucose solution for use in each test tube/using a fixed concentration of glucose solution in each test tube ***[1 mark]***.
d) A control tube should be set up for each pH being investigated, which contains glucose solution but no yeast ***[1 mark]***. No CO_2 should be produced. This will allow the student to check that any CO_2 being released in the other tubes is actually being produced by the yeast ***[1 mark]***.

Topic 5B — Energy Transfer and Nutrient Cycles

Page 19 — Energy Transfer in Ecosystems

1 a) Not all of the energy available from the grass is taken in by the Arctic hare ***[1 mark]***. This is because some parts of the grass aren't eaten, so the energy they contain isn't taken in ***[1 mark]***, and some parts of the grass are indigestible, so they'll pass through the hare and come out as waste ***[1 mark]***. Also, some energy is lost to the environment when the Arctic hare respires ***[1 mark]***.
b) $N = I - (F + R)$
$2345 = 18\,905 - (F + R)$
$F + R = 18\,905 - 2345 =$ **16 560 kJ m^{-2} y^{-1}**
[2 marks for correct answer, otherwise 1 mark for the correct calculation.]

2 a) He could dry out one of his cabbages, e.g. in an oven ***[1 mark]***. He could then burn a known mass of dry tissue in a calorimeter ***[1 mark]*** and use the change in water temperature to calculate the chemical energy stored in the dry biomass of the cabbage ***[1 mark]***.
b) Net primary production, because some of the chemical energy converted by the plant through photosynthesis is immediately used for respiration ***[1 mark]*** so it does not get stored as the biomass of the cabbages ***[1 mark]***.

Answers

Page 21 — Farming Practices and Production

1 a) They eat the crop, reducing the amount of energy available for crop growth ***[1 mark]***.

b) That for the crop shown, the pesticide was most effective at reducing the percentage crop loss to pest 2 ***[1 mark]*** but that it had no effect on reducing the crop loss to the other two pests ***[1 mark]***.

c) Any two from: e.g. use an insecticide that kills multiple pests ***[1 mark]***. / Use another pesticide in conjunction with the first one ***[1 mark]***. / Use biological controls as well as chemical insecticides ***[1 mark]***.

d) Keep them in pens, so respiratory losses through movement are reduced ***[1 mark]***. Keep them warm, so less energy is wasted in generating body heat ***[1 mark]***.

Page 23 — Nutrient Cycles

1 a) A — ammonification ***[1 mark]***, B — nitrogen fixation ***[1 mark]***, C — denitrification ***[1 mark]***

b) i) Saprobionts convert nitrogen compounds in dead organisms, faeces and urine ***[1 mark]*** into ammonia ***[1 mark]***.

ii) They secrete enzymes and digest their food externally ***[1 mark]***, then absorb the nutrients they need ***[1 mark]***.

Page 25 — Fertilisers and Eutrophication

1 a) The control river helps to determine whether it is the fertiliser added to the adjacent field that is causing the observed changes in algal and oxygen content in the river or another variable ***[1 mark]***.

b)
$$\text{percentage change} = \frac{\text{final value} - \text{original value}}{\text{original value}} \times 100$$
$$= \frac{95\ 000 - 10\ 000}{10\ 000} \times 100$$
$= \mathbf{850\%}$ ***[1 mark]***

c) There's a negative correlation between the algal content and the oxygen content of the water / as the algal content increases, the oxygen content decreases, and vice versa ***[1 mark]***.

d) The increasing algal content could have prevented light from reaching plants below ***[1 mark]***, causing them to die and be decomposed by bacteria ***[1 mark]***. The increased numbers of bacteria use up oxygen in the river when carrying out aerobic respiration, resulting in a reduction in dissolved oxygen content ***[1 mark]***. Where algal content is lower, there's less dead plant matter/decomposition and oxygen content is higher ***[1 mark]***.

Extra Exam Practice for Topic 5

Pages 26-27

2.1 Light energy is absorbed by chlorophyll which leads to photoionisation/electrons being released ***[1 mark]***. Some of the energy released from this process is used for the photophosphorylation of ADP/to add a phosphate group to ADP, which generates ATP ***[1 mark]***.

2.2 The constant light would allow the light-dependent reaction of photosynthesis to produce ATP and reduced NADP continuously ***[1 mark]***. These products could then be used continuously by the light-independent reaction/Calvin cycle ***[1 mark]***, to reduce glycerate 3-phosphate/GP to triose phosphate/TP ***[1 mark]***. This would lead to an increase in the production of organic substances, since these are made from triose phosphate/TP ***[1 mark]***.

2.3 Any five from: saprobionts break down the organic compounds in the fish waste ***[1 mark]***. This releases phosphate ions, which can be used by the plants to make molecules containing phosphorus ***[1 mark]***. Saprobionts also turn nitrogen compounds from the fish waste into ammonia ***[1 mark]***, which goes on to form ammonium ions ***[1 mark]***. Nitrifying bacteria then convert the ammonium ions into nitrites ***[1 mark]*** and then nitrates, which are used by the plants to make molecules containing nitrogen ***[1 mark]***.

3.1 Rubisco is an enzyme involved in the Calvin cycle/light-independent reaction of photosynthesis ***[1 mark]***. An increase in rubisco's activity would mean that organic substances could be produced more rapidly ***[1 mark]***. This would result in the producer generating more chemical energy/biomass in a given time, which would increase its *NPP* ***[1 mark]***.

Remember, *NPP* (net primary productivity) is the chemical energy stored in the biomass of a producer in a given area, in a given time, once respiratory losses have been taken into account.

3.2 The efficiency of energy transfer from the producer to primary consumer = 7%

N of primary consumer = (64 375 ÷ 100) × 7
= 4506.25 kJ m^{-2} yr^{-1}

Energy ingested by secondary consumer (*I*) is 90% of 4506.25

I = (4506.25 ÷ 100) × 90 = 4055.625 kJ m^{-2} yr^{-1}

N of secondary consumer = *I* – (*F* + *R*)
= 4055.625 – (2343 + 1261)
= **451.6 kJ m^{-2} yr^{-1}** (to 4 s.f.)

[3 marks for the correct answer, otherwise 1 mark for correctly calculating 'N' of the primary consumer and 1 mark for correctly calculating 'I' of the secondary consumer]

You should know that *F* and *R* both represent ways in which energy is lost — *F* is chemical energy lost in faeces and urine, and *R* is energy lost through respiration.

3.3 By breathing faster, more oxygen will be available in the cells of the secondary consumer ***[1 mark]***. Oxygen is needed as a final electron acceptor in the oxidative phosphorylation stage of aerobic respiration ***[1 mark]***. By increasing the amount of oxygen that is available, oxidative phosphorylation will be able to occur at a faster rate, increasing the rate of ATP synthesis / oxidative phosphorylation will be able to occur, reducing the need for anaerobic respiration, which produces ATP at a slower rate than aerobic respiration ***[1 mark]***.

Topic 6A — Stimuli and Responses

Page 29 — Nervous Communication

1 Receptors detect stimuli ***[1 mark]***. Effectors bring about a response to a stimulus to produce an effect ***[1 mark]***.

2 a) Touch receptors on the surface of the eye (A) are stimulated ***[1 mark]***. An electrical impulse is sent along the sensory neurone (B) to a relay neurone (C) ***[1 mark]***. The impulse is then passed to a motor neurone (D) ***[1 mark]***, which stimulates effector muscles (E) causing them to contract and the person's eyelids to close ***[1 mark]***.

b) Damage to the CNS could interrupt the transmission of the reflex, preventing the reflex response from occurring ***[1 mark]***.

3 Motor neurones carry electrical impulses from the CNS to effectors which then respond ***[1 mark]***. Damage to the motor neurones means the CNS can't communicate with effectors such as muscles ***[1 mark]***, so muscles don't respond and move/are paralysed ***[1 mark]***.

Answers

Page 31 — Responses in Plants and Animals

1 a) The data shows that the plants provided with auxins grew more than those not given auxins ***[1 mark]***. This is because auxins stimulate plant growth (by cell elongation) ***[1 mark]***.

b) Providing tomato plants with auxins could, potentially, be used to increase the height of tomato plants, which might increase the yield of tomatoes/number of tomatoes grown ***[1 mark]***.

c) Auxin is redistributed to the shaded side of the shoot ***[1 mark]***. Auxin stimulates cell elongation on the shaded side ***[1 mark]*** so the shoot bends to grow towards the light ***[1 mark]***.

Page 33 — Receptors

1 When a Pacinian corpuscle is stimulated, the lamellae are deformed and press on the sensory nerve ending ***[1 mark]***. This causes the sensory neurone's cell membrane to stretch and the deformation of stretch-mediated sodium ion channels ***[1 mark]***. The sodium ion channels open and sodium ions diffuse into the cell creating the generator potential ***[1 mark]***.

2 In the retina/fovea, cones are close together and each cone joins one bipolar neurone ***[1 mark]***. When light from two points hits two cones, action potentials from each cone go to the brain ***[1 mark]***. This means you can distinguish two points that are close together as two separate points ***[1 mark]***.

Page 35 — Control of Heart Rate

1 a) The sinoatrial node acts as a pacemaker/sets the rhythm of the heartbeat ***[1 mark]***.

b) The Purkyne tissue conducts electrical impulses through the ventricle walls ***[1 mark]***.

2 a) E.g. chemoreceptors in the aorta/carotid artery/medulla detect the high CO_2 concentration ***[1 mark]***. Impulses are sent from the receptors to the medulla ***[1 mark]***, which sends impulses along sympathetic neurones to the sinoatrial node (SAN) ***[1 mark]***. These neurones secrete noradrenaline, which binds to receptors on the SAN ***[1 mark]***. This increases the SAN activity, which increases heart rate ***[1 mark]***. ***[Up to 3 marks for explaining how impulses get to the SAN, 1 mark for linking increased SAN activity to increased heart rate. Maximum of 4 marks available.]***

b) Low blood O_2 level ***[1 mark]***, low blood pH level ***[1 mark]***.
The low blood pH level is caused by the increased CO_2 level.

3 a) The AVN passes waves of electrical activity on to the bundle of His and the Purkyne tissue to make the ventricles contract ***[1 mark]***. By stopping the AVN from functioning, the rapid irregular impulses from the atria aren't transmitted via the bundle of His and the Purkyne tissue to the ventricles, so they can't affect the heart rate (i.e. make it high and/or irregular) ***[1 mark]***.

b) Without a functioning AVN the heart can't beat normally/ventricles can't contract normally ***[1 mark]***. A pacemaker is needed to generate electrical impulses that cause the heart to beat normally/ventricles to contract normally ***[1 mark]***.

Topic 6B — Nervous Coordination

Page 38 — Neurones

1 a) A stimulus causes sodium ion channels in the neurone cell membrane to open ***[1 mark]***. Sodium ions diffuse into the cell, so the membrane becomes depolarised ***[1 mark]***.

b) The first action potential fired at 0.5 ms. If the second one fired at 4.5 ms, this means an action potential is fired every (4.5 – 0.5 =) 4 ms.
Number of ms in one hour = 60 × 60 × 1000 = 3 600 000.
There is one action potential every 4 ms, so in one hour there will be 3 600 000 ÷ 4 = 900 000 = $\mathbf{9 \times 10^5}$ action potentials.
[2 marks for the correct answer, allow 1 mark for the correct calculation of 3 600 000 ÷ 4.]
There's a lot to do to get the marks here, but that's A-level Biology for you. Just take your time and make sure you write down your calculations — that way you might pick up a mark even if you don't get the final answer right.

c) 30 mV ***[1 mark]***
This is the same as the maximum potential difference shown on the graph. Remember, action potentials always fire with the same change in voltage no matter how big the stimulus is.

Page 41 — Synaptic Transmission

1 a) It is the threshold that needs to be reached for an action potential to fire ***[1 mark]***.

b) Any four from: before the action potential fired, the potential difference across the membrane increased three times in quick succession ***[1 mark]***. The increases in potential difference were caused by nerve impulses arriving at the synapse and releasing neurotransmitter ***[1 mark]***, which caused sodium ion channels to open on the postsynaptic membrane ***[1 mark]***. This allowed an influx of sodium ions into the postsynaptic membrane, which increased the potential difference across the membrane ***[1 mark]***. It was not until the arrival of the third impulse that enough neurotransmitter was acting on the membrane to allow the threshold level to be reached and the action potential to be fired ***[1 mark]***. ***[Maximum of 4 marks available.]***

2 There will be fewer receptors for acetylcholine/ACh to bind to ***[1 mark]***, so fewer sodium ion channels will open at neuromuscular junctions ***[1 mark]***, making it less likely that action potentials will be generated in the muscle cells ***[1 mark]***.

3 Galantamine would stop acetylcholinesterase/AChE breaking down acetylcholine/ACh, so there would be more acetylcholine/ACh in the synaptic cleft ***[1 mark]*** and it would be there for longer ***[1 mark]***. This means more nicotinic cholinergic receptors would be stimulated ***[1 mark]***.

Page 43 — Muscle Contraction

1 a) A = sarcomere ***[1 mark]***.
B = Z-line ***[1 mark]***.
C = H-zone ***[1 mark]***.

b) The A-bands stay the same length during contraction ***[1 mark]***. The I-bands get shorter ***[1 mark]***.

c) Drawing number 3 ***[1 mark]*** because the M-line connects the middle of the myosin filaments ***[1 mark]***. The cross-section would only show myosin filaments, which are the thick filaments ***[1 mark]***.
The answer isn't drawing number 1 because all the dots in the cross-section are smaller, so the filaments shown are thin actin filaments — which aren't found at the M-line.

Page 45 — Muscle Contraction

1 Muscles need ATP to relax because ATP provides the energy to break the actin-myosin cross bridges ***[1 mark]***. If the cross bridges can't be broken, the myosin heads will remain attached to the actin filaments ***[1 mark]***, so the actin filaments can't slide back to their relaxed position so the muscle stays contracted ***[1 mark]***.

2 The muscles won't contract ***[1 mark]*** because calcium ions won't be released into the sarcoplasm, so tropomyosin will continue to block the actin-myosin binding sites ***[1 mark]***. This means no actin-myosin cross bridges can be formed ***[1 mark]***.

Topic 6C — Homeostasis

Page 47 — Homeostasis Basics

1 a) Statement A because body temperature continues to increase from the normal level and isn't returned ***[1 mark]***.

b) It makes metabolic reactions less efficient ***[1 mark]*** because the enzymes that control metabolic reactions may denature ***[1 mark]***.

2 Multiple negative feedback mechanisms give more control over changes in the internal environment than just having one feedback mechanism ***[1 mark]***. This is because you can actively increase or decrease a level so it returns to normal ***[1 mark]***.

Answers

Page 49 — Control of Blood Glucose Concentration

1 a) Negative feedback because the pancreas secretes hormones that return blood glucose concentration to normal if it is detected as being too high or too low ***[1 mark]***.

b) Insulin binds to specific receptors on muscle cells causing them to become more permeable to glucose, so more is absorbed from the blood ***[1 mark]***. / Insulin activates glycogenesis, so that glucose can be stored as glycogen ***[1 mark]***. / Insulin causes the rate of respiration of glucose to increase, so that more glucose is used up ***[1 mark]***.

2 When adrenaline and glucagon bind to the receptors on the cell membrane they activate an enzyme called adenylate cyclase ***[1 mark]***. Activated adenylate cyclase converts ATP into cAMP, a second messenger ***[1 mark]***. cAMP activates protein kinase A, which activates a cascade that breaks down glycogen into glucose ***[1 mark]***.

Page 51 — Control of Blood Glucose Concentration

1 a) Any two from: Person A's blood glucose concentration is initially at a higher level than person B's blood glucose concentration ***[1 mark]***. / Person A's blood glucose concentration reaches a much higher level than person B's blood glucose concentration ***[1 mark]***. / It takes longer for person A's blood glucose concentration to start to decrease than it does for person B's blood glucose concentration to start to decrease ***[1 mark]***. / Person A's blood glucose concentration decreases at a much slower rate than person B's blood glucose concentration ***[1 mark]***. ***[Maximum of 2 marks available.]***

b) The insulin receptors on person A's cell membranes don't work properly, so the cells don't take up enough glucose ***[1 mark]***. This means their blood glucose concentration remains higher than normal ***[1 mark]***.

Page 53 — The Kidneys

1 a) The efferent arteriole has a smaller diameter than the afferent arteriole, so the blood in the glomerulus is under high pressure ***[1 mark]***. The high pressure forces liquid and small molecules into the Bowman's capsule (point A), forming the glomerular filtrate ***[1 mark]***.

b) Point C, because glucose is reabsorbed in the proximal convoluted tubule/PCT, so by the time the filtrate reaches point C there will be less glucose remaining ***[1 mark]***.

c) If there is 0 mg of glucose in the urine, all the glucose filtered out of the blood must be reabsorbed. So:
$6300 \times 0.9 = 5670$ mg hour^{-1}
$5670 \div 60 =$ **94.5 mg min^{-1}** ***[1 mark]***

Page 55 — Controlling Blood Water Potential

1 a) Strenuous exercise causes more sweating, so more water is lost ***[1 mark]***. This decreases the water potential of the blood ***[1 mark]***. This is detected by osmoreceptors in the hypothalamus ***[1 mark]***, which stimulates the posterior pituitary gland to release more ADH ***[1 mark]***.

b) The ADH increases the permeability of the walls of the distal convoluted tubule and collecting duct ***[1 mark]***. This means more water is reabsorbed into the medulla and into the blood by osmosis ***[1 mark]***.

2 A longer descending limb, means more water can be reabsorbed into the blood from the nephron in the descending limb ***[1 mark]***. A longer ascending limb means more ions are actively pumped out into the medulla ***[1 mark]***, which creates a really low water potential in the medulla ***[1 mark]***. This means more water moves out of the collecting duct into the capillaries, giving a low volume of urine ***[1 mark]***.

Extra Exam Practice for Topic 6

Pages 56-57

2.1 Each cone cell in the eye connects to the optic nerve with one neurone, whereas one neurone connects several rod cells to the optic nerve ***[1 mark]***. This means that cone cells give higher visual acuity/clearer vision than rod cells ***[1 mark]*** because information about light hitting each cone cell can be interpreted separately by the brain, whereas information about light hitting each rod cell is combined before it reaches the brain ***[1 mark]***.

2.2 If glucose transporter proteins are blocked it means that glucose can't be reabsorbed into the blood from the kidneys ***[1 mark]***. The unabsorbed glucose will remain in the glomerular filtrate, meaning the glucose concentration of the urine will increase ***[1 mark]***.

2.3 Absorbance = 0.5 ± 0.05
$= 0.45 - 0.55$
Range of glucose concentration = **2.2-3.2 mM**
[2 marks for the correct answer, otherwise 1 mark for correctly calculating the range of absorbance values]
The uncertainty is in the absorbance of the colorimeter, so firstly you need to work out the range of absorbance values that would give a reading of 0.5. Then read off the glucose concentrations at these two absorbance values to get the range in glucose concentration.

3.1 A phosphate group taken from phosphocreatine can be added to ADP to make ATP ***[1 mark]***, to provide the energy needed for muscle contraction ***[1 mark]***.

3.2 The calcium ions leave their binding site on the protein attached to tropomyosin ***[1 mark]***. This causes the tropomyosin molecules to move back so they block the myosin head binding sites on actin again ***[1 mark]***. Because there are no actin-myosin cross-bridges, the actin filaments slide back to their relaxed position, lengthening the sarcomere ***[1 mark]***.

3.3 A person with diabetes insipidus would be more at risk of dehydration, because not as much ADH would be released when the fall in blood water potential was detected ***[1 mark]***. This would mean that the permeability of the DCT and collecting duct wouldn't be increased as much as normal ***[1 mark]***, so not as much water would be reabsorbed into the blood by osmosis from the glomerular filtrate ***[1 mark]***. With less water being reabsorbed into the blood, it would be more difficult to increase the blood water potential, meaning the person would be more at risk of dehydration ***[1 mark]***.

Topic 7A — Genetics

Page 60 — Inheritance

1 Parents' genotypes identified as RR and rr ***[1 mark]***. Correct genetic diagram drawn with gametes' alleles identified as R, R and r, r ***[1 mark]*** and gametes crossed to show Rr as the only possible genotype in the offspring ***[1 mark]***.
The question specifically asks you to draw a genetic diagram so make sure that you include one in your answer, e.g.

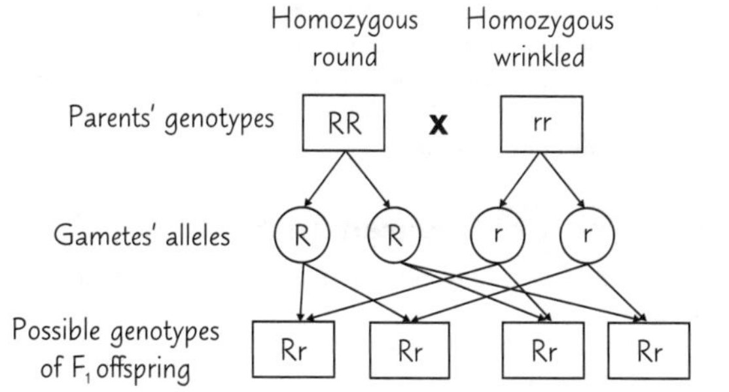

Answers

2 a) Because the alleles for red and white coats are codominant, so they are both expressed in the phenotype ***[1 mark]***.

b) Parents' genotypes identified as C^WC^W and C^RC^W ***[1 mark]***. Correct genetic diagram drawn with gametes' alleles identified as C^W, C^W and C^R, C^W ***[1 mark]*** and gametes crossed to show two offspring with genotype C^WC^W and two with genotype C^RC^W ***[1 mark]***. The phenotypes of the offspring are stated as two white and two roan ***[1 mark]***.

The question specifically asks you to draw a genetic diagram so make sure that you include one in your answer, e.g.

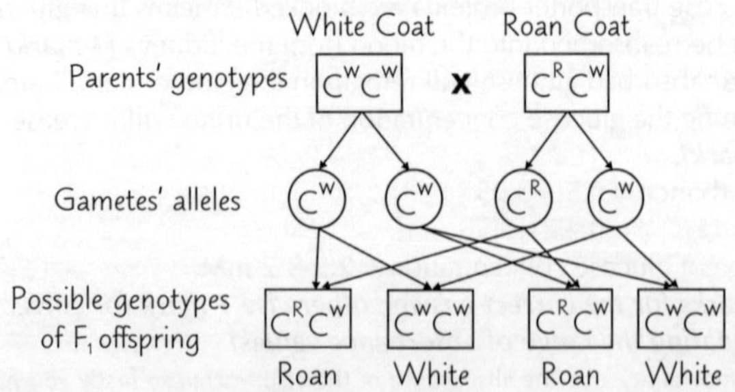

Page 63 — Linkage and Epistasis

1 a) Parents' genotypes identified as X^HX^h and X^hY ***[1 mark]***. Correct genetic diagram drawn with gametes' alleles identified as X^H, X^h and X^h,Y ***[1 mark]*** and gametes crossed to show X^HX^h, X^HY, X^hX^h and X^hY as the possible genotypes of the offspring ***[1 mark]***.

The question specifically asks you to draw a genetic diagram, so make sure that you include one in your answer, e.g.

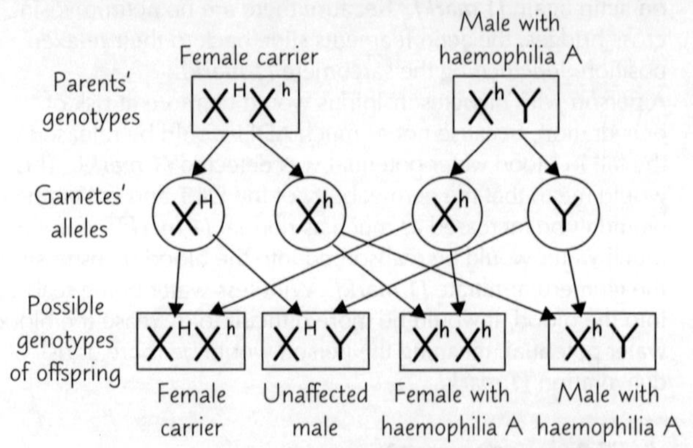

b) Men only have one copy of the X chromosome (XY) but women have two (XX) ***[1 mark]***. Haemophilia A is caused by a recessive allele, so females would need two copies of the allele for them to have haemophilia A ***[1 mark]***. As males only have one X chromosome they only need one recessive allele to have haemophilia A, which makes them more likely to have haemophilia A than females ***[1 mark]***.

2 The table shows that a cross between hhss and HHSS produces a 36 : 9 : 3 or 12 : 3 : 1 phenotypic ratio in the F_2 generation of bald : straight hair : curly hair ***[1 mark]***. This is because the hair gene has a dominant epistatic allele (H) ***[1 mark]***, which means having at least one copy of the dominant epistatic gene (Hh or HH) will result in a bald phenotype that masks the expression of the type of hair gene ***[1 mark]***.

Page 65 — The Chi-Squared Test

1 a) There's no difference between the observed and expected results ***[1 mark]***.

b) Yes, the data supports the theory that petal colour in the flower is controlled by dihybrid inheritance. The χ^2 value is smaller than the critical value ***[1 mark]*** so the scientist is unable to reject the null hypothesis ***[1 mark]***.

Topic 7B — Populations and Evolution

Page 67 — The Hardy-Weinberg Principle

1 a) Frequency of genotype TT = p^2 = 0.14
So the frequency of the dominant allele = p = $\sqrt{0.14}$ = 0.37
The frequency of the recessive allele = q
q = 1 – p
q = 1 – 0.37 = **0.63** ***[2 marks for the correct answer or 1 mark for 1 – √0.14]***

b) Frequency of homozygous recessive genotype tt = q^2 = 0.63^2 = **0.40** ***[1 mark. Allow 1 mark for evidence of correct calculation using incorrect answer to part a]***.

c) Those that don't have a cleft chin are homozygous recessive tt = 40%, so the percentage that do have a cleft chin, Tt or TT, is 100% – 40% = **60%** ***[1 mark]***.

There are other ways of calculating this answer, e.g. working out the value of 2pq and adding it to p^2. It doesn't matter which way you do it as long as you get the right answer.

2 Frequency of allele F^B = 43% = 0.43
So the frequency of allele F^W = 1 – 0.43 = 0.57
The frequency of the heterozygous genotype = 2pq
2pq = 2(0.43 × 0.57) = **0.49** ***[2 marks for the correct answer or 1 mark for 2(0.43 × 0.57).]***

In the Hardy-Weinberg equations, 'p' is usually the dominant allele and 'q' is usually the recessive allele, but this doesn't have to be the case. In this scenario, there's no recessive allele, so you can just make 'p' represent one of the alleles and 'q' represent the other — it doesn't matter which way round you do it either. If you're not told which is the dominant and which is the recessive allele in an exam question, you can do the same thing.

Page 69 — Variation and Selection

1 a) As temperature decreases from 22 °C to 16 °C the frequency of h, the long hair allele, increases from 0.11 to 0.23 ***[1 mark]***. This could be because the allele for long hair is more beneficial at colder temperatures ***[1 mark]***. Hamsters with the h allele will have a greater chance of surviving, reproducing and passing on their genes, including the beneficial h allele ***[1 mark]***. So a greater proportion of the next generation will inherit the beneficial allele and the frequency of the h allele will increase ***[1 mark]***.

b) Directional selection ***[1 mark]***.

Page 71 — Speciation and Genetic Drift

1 a) E.g. The new species could not breed with each other ***[1 mark]***.

b) Different populations of flies were physically/geographically isolated and experienced different selection pressures (different food) ***[1 mark]***. This led to changes in allele frequencies between the populations ***[1 mark]***, which made them reproductively isolated/unable to interbreed and produce fertile offspring, and eventually resulted in speciation ***[1 mark]***.

Topic 7C — Populations in Ecosystems

Page 73 — Ecosystems

1 Their wings are light and flexible, which allows them to catch fast and manoeuvrable insects. This increases their chances of catching enough food to survive ***[1 mark]***. They use echolocation so they can catch insects that come out at night. This also increases their chances of catching enough food to survive ***[1 mark]***. They make unique mating calls so they only attract a mate of the same species. This increases their chance of reproduction by making successful mating more likely ***[1 mark]***.

This question is only asking about the biotic conditions (the living features of the ecosystem), so you won't get any marks for talking about abiotic conditions (the non-living features of the ecosystem).

Answers

Page 75 — Variation in Population Size

1 a) $\text{Rate} = \frac{\text{Change in } y}{\text{Change in } x} = \frac{25}{4}$

$= \mathbf{6.25\ thousand\ year^{-1}\ /\ 6250\ year^{-1}}$

[2 marks for the correct answer or 1 mark for the correct calculation.]

'Year^{-1}' means 'per year'.

b) The predator population increased as the prey population increased because there was more food available for the predators ***[1 mark]***. The population of prey then fell because many prey were eaten by the large population of predators ***[1 mark]***. The predator population then fell because there was less prey for the predators to eat ***[1 mark]***.

c) Between 25 and 30 years ***[1 mark]***. The prey population starts to decline but predator numbers have been very low for several years, suggesting that the prey are competing with one another for space and food / space and food have become the limiting factors for the prey population size ***[1 mark]***.

Page 77 — Investigating Populations

1 a) E.g. the field could be divided into a grid a random number generator could be used to select random coordinates on the grid ***[1 mark]***. frame quadrats could be placed on the ground at these random coordinates ***[1 mark]***. The percentage of each frame quadrat that's covered by clover plants could be recorded ***[1 mark]***. The percentage cover for the whole field could then be estimated by taking a mean of the data collected in all of the frame quadrats ***[1 mark]***.

Clover plants are small and grow very close together, so it's much easier to estimate their population size using percentage cover, rather than trying to count individual plants.

b) E.g. including plant species that aren't clover plants could increase the estimate of percentage cover / ignoring clover plants could reduce the estimate of percentage cover ***[1 mark]***.

Page 79 — Succession

1 a) Primary succession ***[1 mark]*** because there is no soil or organic matter ***[1 mark]***.

b) When the grass dies, microorganisms decompose the dead organic material, forming a soil ***[1 mark]***. The formation of soil helps to retain water and makes the conditions less hostile, which allows larger plants, like shrubs, to move in ***[1 mark]***.

Page 81 — Conservation

1 a) There's a link between fishing mortality rate and the cod stock size ***[1 mark]***. As the fishing mortality rate increases, the cod stock size decreases/there's a negative correlation between fishing mortality rate and cod stock size ***[1 mark]***.

b) E.g. it could be used by governments to make decisions about cod fishing quotas (the amount of cod allowed to be removed from the sea by fishermen each year) ***[1 mark]***.

c) The conservationists will want to limit the amount of fishing to a sustainable level to maintain fish stocks for future generations ***[1 mark]***. However, limiting the amount of fishing may reduce the incomes of people employed in the fishing industry ***[1 mark]***.

2 a) It provides wood for people to use whilst preserving some trees which can continue to grow and provide wood in the future ***[1 mark]***.

b) Any two from: e.g. it maintains the woodland habitat for other organisms ***[1 mark]***. / It allows new trees to grow from seeds produced by the mature standards ***[1 mark]***. / The mature standard can be used to produce larger logs at a later date ***[1 mark]***.

c) The canopy of mature standard trees will block out the light that the coppiced trees need to grow ***[1 mark]***.

Extra Exam Practice for Topic 7

Pages 82-83

2.1 Primary succession occurs when pioneer species colonise new land surfaces that don't contain soil ***[1 mark]***. As the grasses in Figure 2 require soil and begin growing from time = 0, soil must already be in the environment and so Figure 2 cannot be showing primary succession ***[1 mark]***.

2.2 Removing trees at point A would prevent succession ***[1 mark]*** and therefore maintain more shrubs and grasses in the woodland, which are likely to be important habitats ***[1 mark]***.

2.3 E.g. at point C the biomass of trees is greater than it is at point B, which suggests that the shrub species face greater competition from trees at point C ***[1 mark]***. Greater competition is likely to reduce the carrying capacity of shrub species, as there are not enough resources to maintain the population size of each species ***[1 mark]***.

Remember, the carrying capacity is the maximum stable population size of a species that an ecosystem can support.

2.4 Directional selection is likely to have acted on the trees ***[1 mark]***. Trees with the allele(s) for thicker bark would have been more likely to survive wildfires and to reproduce, passing on the allele(s) for thicker bark ***[1 mark]***. Over time this increased the frequency of the allele(s) for thicker bark in the population of trees frequently exposed to wildfires ***[1 mark]***.

2.5 $$\text{Total population size} = \frac{\text{Number caught in first sample} \times \text{Number caught in second sample}}{\text{Number marked in second sample}}$$

Number caught in first sample after wildfire = 28
Number caught in second sample after wildfire = 24
Number marked in second sample after wildfire = 12
Total population size after wildfire = (28 × 24) ÷ 12
= 56 individuals

$$\%\ \text{change} = \frac{\text{population size after wildfire} - \text{population size before wildfire}}{\text{Population size before wildfire}} \times 100$$

$$= \frac{56 - 336}{336} \times 100 = \frac{-280}{336} \times 100$$

$= \mathbf{-83\%}$ (to 2 s.f.)

[3 marks for the correct answer, otherwise 1 mark for evidence of the correct formula for calculating population size and 1 mark for 56 individuals after wildfire]

The population size has decreased, so make sure you include a minus sign in your answer or else clearly say that your answer is a percentage decrease.

Topic 8A — Mutations and Gene Expression

Page 85 — Mutations

1 a) AG<u>G</u>TAT<u>G</u>AGGC<u>C</u> ***[1 mark]***.

b) The original gene codes for the amino acid sequence serine-tyrosine-glutamine-alanine and the mutated gene codes for the amino acid sequence arginine-tyrosine-glutamic acid-alanine ***[1 mark]***. Even though there are three mutations, there are only two changes to the amino acid sequence ***[1 mark]***. This is because of the degenerate nature of the DNA code, which means more than one codon can code for the same amino acid ***[1 mark]***. So the substitution mutation on the last triplet doesn't alter the amino acid (GCT and GCC both code for alanine) ***[1 mark]***.

Answers

Page 87 — Cancer

1 a) Any two points from: e.g. malignant tumours are cancers. Benign tumours are not cancerous ***[1 mark]***. Malignant tumours usually grow rapidly. Benign tumours usually grow slower than malignant tumours ***[1 mark]***. Malignant tumours can invade and destroy surrounding tissues/spread to other parts of the body. Benign tumours can't ***[1 mark]***.

b) If a mutation occurs in a tumour suppressor gene ***[1 mark]***, proteins that stop cells dividing and cause cell death might not be produced ***[1 mark]***. If a mutation occurs in a proto-oncogene ***[1 mark]***, it can turn it into an oncogene (an overactive version of the proto-oncogene) causing the production of too many proteins that cause cells to divide ***[1 mark]***. In both cases, the mutation allows cells to grow and divide uncontrollably ***[1 mark]***.

c) Oestrogen can stimulate some breast cells to divide and replicate ***[1 mark]***. Because more replication is taking place, the chances of new cancer-causing mutations being introduced increases ***[1 mark]***. This stimulation could also help already cancerous cells replicate ***[1 mark]***. Some research also suggests that oestrogen can directly cause mutations in certain breast cells, which again increases the chance of cancer-causing mutations being introduced ***[1 mark]***.

Page 89 — Interpreting Data on Cancer

1 E.g. if the screen revealed that a woman had the BRCA1 mutation, she could be screened for signs of breast cancer more regularly than the rest of the population, so the cancer could be diagnosed early if it does develop ***[1 mark]***. She would also be aware that she had a higher risk of developing breast cancer, so would know to be more vigilant when checking for signs of the disease ***[1 mark]***. She could also choose to take steps to reduce the risk developing breast cancer, such as having a mastectomy ***[1 mark]***. If the disease did develop, knowing the mutation that has caused it could also help to determine the specific treatment used to give the best chance of survival ***[1 mark]***.

Page 93 — Stem Cells

1 a) E.g. stem cell therapies are currently being used for some diseases affecting the blood and immune system ***[1 mark]***. Bone marrow contains stem cells that can become specialised to form any type of blood cell ***[1 mark]***. Bone marrow transplants can be used to replace faulty bone marrow in patients with leukaemia (a cancer of the blood or bone marrow) ***[1 mark]***. The stem cells in the transplanted bone marrow divide and specialise to produce healthy blood cells ***[1 mark]***.

b) Obtaining embryonic stem cells involves the destruction of an embryo ***[1 mark]***. Some people believe that embryos have a right to life and that it's wrong to destroy them ***[1 mark]***.

c) E.g. induced pluripotent stem cells are produced by 'reprogramming' specialised adult body cells to become pluripotent ***[1 mark]***. To do this, the adult body cells are made to express a series of transcription factors that are normally associated with pluripotent stem cells ***[1 mark]***. The genes that code for the transcription factors are introduced to the adult cell's DNA ***[1 mark]*** using a modified virus that has the genes within its own DNA ***[1 mark]***.

Page 96 — Regulation of Transcription and Translation

1 a) The results of tubes 1 and 2 suggest that oestrogen affects the expression of the gene for the Chi protein ***[1 mark]*** because mRNA and active protein production only occur in the presence of oestrogen ***[1 mark]***.

b) The mutant could have a faulty oestrogen receptor ***[1 mark]***. Oestrogen might not bind to the receptor / the oestrogen-oestrogen receptor complex might not work as an activator ***[1 mark]***. This would mean even in the presence of oestrogen transcription wouldn't be activated, so no mRNA or protein would be produced ***[1 mark]***.

This is a pretty tricky question — drawing a diagram of how oestrogen controls transcription would help you figure out the answer.

c) E.g. the siRNA and associated proteins would attach to the mRNA of the Chi protein and cut it up into smaller portions ***[1 mark]***, resulting in no full length mRNA ***[1 mark]***. No mRNA would be available for translation, so no protein would be produced ***[1 mark]***.

Page 98 — Epigenetic Control of Gene Expression

1 a) Histones are proteins that DNA wraps around to form chromatin, which makes up chromosomes ***[1 mark]***.

b) When acetyl groups are removed from the histones in chromatin, the chromatin becomes highly condensed ***[1 mark]***. This means that the enzymes/proteins needed for transcription cannot access the DNA ***[1 mark]*** and the DNA cannot be transcribed ***[1 mark]***.

c) E.g. acetyl groups are removed from histones by histone deacetylase (HDAC) enzymes ***[1 mark]***. Drugs can be used to inhibit these enzymes ***[1 mark]***. This means that the histones remain acetylated and the DNA associated with them can be transcribed as normal ***[1 mark]***.

Page 99 — Evaluating Data on Phenotypes

1 That genetic factors have a bigger influence than environmental factors on stuttering ***[1 mark]***.

Topic 8B — Genome Projects and Gene Technologies

Page 102 — Genome Projects and Making DNA Fragments

1 a) There's a BamHI recognition sequence at either side of the DNA fragment, so you could use this restriction endonuclease to isolate the fragment ***[1 mark]***. BamHI would be incubated with the bacterial DNA, so that it cuts the DNA at each of these recognition sequences ***[1 mark]***.

b) The genetic code is universal/all organisms use the same genetic code ***[1 mark]***. Transcription and translation mechanisms are similar in different species ***[1 mark]***.

c) Simple organisms, like bacteria, have fewer non-coding regions than more complex organisms such as plants ***[1 mark]***. Plants also have regulatory genes and bacteria don't ***[1 mark]***. This makes it harder to find the parts that code for proteins in the plant's DNA than in the bacteria's DNA ***[1 mark]***.

You're effectively being asked to compare the difficulty of translating the genome into the proteome for two different organisms here — when answering any comparison question, make sure you talk about both of the things you're comparing in your answer.

d) mRNA that's complementary to the DNA fragment is isolated from the cells ***[1 mark]*** and mixed with free DNA nucleotides and reverse transcriptase ***[1 mark]***. The reverse transcriptase uses the mRNA as a template to synthesise a new strand of cDNA ***[1 mark]***.

Page 104 — Amplifying DNA Fragments

1 a) Colony A is visible/fluoresces under UV light, but Colony B isn't visible/doesn't fluoresce ***[1 mark]***. So only Colony A contains the fluorescent marker gene, which means it contains transformed cells ***[1 mark]***.

Answers

b) The plasmid vector DNA would have been cut open with the same restriction endonuclease that was used to isolate the DNA fragment containing the target gene ***[1 mark]***. The plasmid DNA and gene (DNA fragment) would have been mixed together with DNA ligase ***[1 mark]***. DNA ligase joins the sticky ends of the DNA fragment to the sticky ends of the plasmid DNA ***[1 mark]***.

Page 107 — Using Recombinant DNA Technology

1 a) The drought-resistance gene could be inserted into a plasmid ***[1 mark]***. The plasmid is then inserted into a bacterium ***[1 mark]***, which is used as a vector to get the gene into the plant cells ***[1 mark]***.

b) The transformed wheat plants could be grown in drought-prone regions ***[1 mark]***, where they would reduce the risk of famine and malnutrition ***[1 mark]***.

c) They could be concerned that the large agricultural company will have control over the recombinant DNA technology used to make the drought-resistant plants, which could force smaller companies out of business ***[1 mark]***.

Page 109 — Gene Probes and Medical Diagnosis

1 a) The allele that you want to screen for is sequenced ***[1 mark]***. Multiple complementary copies of parts of the allele are made by PCR to be used as DNA probes ***[1 mark]***.

b) Microscopic spots of different DNA probes are attached in series to a glass slide, producing a microarray ***[1 mark]***. A sample of the person's labelled DNA is washed over the array and if any of the DNA matches any of the probes, it will stick to the array ***[1 mark]***. The array is washed and visualised, under UV light/X-ray film ***[1 mark]***. Any spot that shows up means that the person's DNA contains that specific allele ***[1 mark]***.

2 a) Because the patient tested negative for the mutated allele (KRAS oncogene) that the drug specifically targets ***[1 mark]***.

b) So the results of the patient's screening can be explained to them ***[1 mark]*** and so the treatment options can also be explained ***[1 mark]***.

Page 111 — Genetic Fingerprinting

1 a) Genetic fingerprinting is based on comparing the length of variable number tandem repeats/VNTRs at particular points on the genome ***[1 mark]***. PCR is used to make copies of/amplify the areas of DNA that contain the VNTRs ***[1 mark]***. This produces many DNA fragments for analysis with gel electrophoresis, which produces a genetic fingerprint ***[1 mark]***.

b) Genetic fingerprint 1 is most likely to be from the child's father because five out of six of the bands on his genetic fingerprint match that of the child's, compared to only one on fingerprint 2 ***[1 mark]***.

c) Any two from: e.g. it can be used to link a person to a crime scene (forensic science). / To prevent inbreeding between animals or plants. / To diagnose cancer or genetic disorders. / To investigate the genetic variability of a population. ***[1 mark for two correct answers.]***

Extra Exam Practice for Topic 8

Pages 112-113

2.1 Multipotent because they can produce different types of cell, whereas unipotent cells can only become one type of cell ***[1 mark]***.

2.2 Transcription factors bind to the DNA in the cell nucleus ***[1 mark]***. These control which genes are expressed ***[1 mark]***, so that only the genes that code for proteins specific for red blood cells are transcribed and translated ***[1 mark]***. These proteins modify the cell, resulting in red blood cells ***[1 mark]***.

2.3 The stem cells that are transplanted would need to be haematopoietic stem cells, since these are the cells which develop into the cells that are affected by MDS ***[1 mark]***. As the defect that causes MDS is found in haematopoietic stem cells, a donor would be needed to obtain healthy versions of these cells ***[1 mark]***.

2.4 Metabolised azacitidine binds irreversibly to DNA methyltransferases, so there will be a reduced amount of methylation (at CpG sites) on tumour suppressor genes ***[1 mark]***. As a result, tumour suppressor genes are expressed ***[1 mark]*** and so proteins that stop cells from dividing / cause cells to self-destruct/apoptosis are produced ***[1 mark]***. This means that the rate of cell division/mitosis is controlled ***[1 mark]***.

2.5 Azacitidine counteracts the epigenetic change (abnormal methylation of tumour suppressor genes) that causes the abnormal cell growth seen in MDS ***[1 mark]***.

2.6 E.g. yes, as for three out of the five studies shown in Table 1 ***[1 mark]*** the odds ratio is lower for former smokers than it is for current smokers ***[1 mark]***.

Do Well In Your Exams

Page 122 — How to Do Well in Your Exams

Q1 ***21-25 marks:***
The answer includes material from a variety of different topic areas and clearly shows its link to the question title.
No irrelevant material is included. The answer includes a range of detailed and accurate biological facts that are all of A-level standard. No incorrect material is included. Appropriate scientific terminology is used. Explanations are clear and the overall essay is very well written.
(To get top marks, evidence of wider reading beyond the specification must be shown.)

16-20 marks:
The answer includes material from several relevant topic areas and links these to the question title. An irrelevant topic may be included. The answer includes a range of biological facts that are accurate and of A-level standard but may sometimes be lacking in detail. There may be one significant error in the scientific content. Appropriate scientific terminology is used. Explanations are clear.

11-15 marks:
The answer includes material from several relevant topic areas but doesn't link them to the question title. More than one irrelevant topic may be included. The biological facts included in the answer are mostly correct and of A-level standard but material is lacking in detail. There may be a few significant errors in the scientific content. Appropriate scientific terminology is usually used. Explanations are usually clear.

6-10 marks:
The answer includes material from one or two relevant topic areas but doesn't link them to the question title.
Several irrelevant topic areas may be included. Some A-level content may be included but it will be lacking in detail and may contain several significant scientific errors. There may be limited use of scientific terminology. Explanations lack clarity.

1-5 marks:
The answer includes material that is only vaguely linked to the question title. Material is presented as a series of facts.
Most of the material is irrelevant. The content is below A-level standard and contains a large number of scientific errors.
Scientific terminology is not used or is below A-level standard.
Explanations are poor or absent.

0 marks:
Nothing relevant is included in the answer or nothing has been written.

Here are some topic areas you might write about:

- enzymes catalysing important cellular reactions (e.g. in photosynthesis and respiration);
- carrier and co-transport proteins aiding facilitated diffusion and active transport of materials across cell membranes;

Answers

- antigens and antibodies in the immune response against pathogens;
- proteins producing beneficial phenotypes in natural selection;
- protein ion channels in cell membranes allowing action potentials to be generated and nervous responses to stimuli to take place;
- the role of actin and myosin proteins in muscle contraction and movement;
- receptor proteins on the surface of cells allowing hormonal responses to stimuli to take place (e.g. insulin and glucagon receptors in the control of blood glucose concentration);
- proteins produced by tumour suppressor genes and proto-oncogenes controlling cell division.

This is not a full list of all the topic areas you could write about — it's just to give you an idea. Remember, you should aim to write about at least five of these topic areas. Whatever topic areas you include, you must relate them to the essay title — so in this case, don't just write about proteins, make it really clear how proteins are important to living organisms.

Synoptic Practice

Pages 123-127

1.1 Shivering is likely to increase the rate of glycolysis as the muscles need more ATP for contraction ***[1 mark]***, which is produced during glycolysis ***[1 mark]***.

1.2 As the muscles use more oxygen, the oxygen concentration in the blood may fall ***[1 mark]***, which would be detected by chemoreceptors in the aorta/carotid artery/medulla ***[1 mark]***. Impulses would be sent from the chemoreceptors to the medulla ***[1 mark]***, which would send impulses along the sympathetic neurones ***[1 mark]***. These would secrete noradrenaline onto receptors of the SAN in the heart, causing heart rate to increase ***[1 mark]***.

1.3 Standard deviation is better for judging the precision of the results as this shows the spread of data around the mean ***[1 mark]***, whereas the range just shows the total spread of the data ***[1 mark]***.

1.4 $$\text{Percentage difference} = \frac{\text{final value} - \text{original value}}{\text{original value}} \times 100$$

$1.2 - 0.3 = 0.9$

$(0.9 \div 0.3) \times 100 =$ **300%** ***[1 mark]***

2.1 The drug will prevent acetylcholine from binding to receptors on the postsynaptic membrane/motor end plate, meaning that sodium ion channels will not open ***[1 mark]***. As there is no influx of sodium ions, the postsynaptic membrane/motor end plate will not become depolarised ***[1 mark]*** and an action potential will not be generated in the target/muscle cells ***[1 mark]***. This will mean that the sarcoplasmic reticulum will not be triggered to release calcium ions ***[1 mark]*** and so the actin-myosin binding sites will remain blocked by tropomyosin, preventing muscle contraction ***[1 mark]***.

2.2 The drug prevents the membrane from repolarising / returning to its resting potential ***[1 mark]***, which prevents the generation of a new action potential and any further muscle contraction ***[1 mark]***.

2.3 E.g. it could prevent the influx of calcium ions into the presynaptic neurones so the vesicles containing acetylcholine are not triggered to release it. / It could increase the rate of acetylcholine breakdown in the synaptic cleft, so that it can't bind to receptors on the postsynaptic membrane/motor end plate and generate an action potential. / It could bind to acetylcholine in the synaptic cleft to prevent it from binding to receptors on the postsynaptic membrane/motor end plate and generating an action potential. ***[1 mark]***

2.4 Minimum dose at which no more than 15% of usual force of contraction remaining (the mean reduction in force of contraction is no less than 85%) = 200 μg kg body mass^{-1}.

$200 \times 75 = 15\,000$ μg

15 000 μg ÷ 1000 = **15 mg**

[2 marks for the correct answer, otherwise 1 mark for 15 000 μg or for a correct conversion from μg to mg.]

2.5 A Student's t-test ***[1 mark]*** because mean values for two data sets are being compared ***[1 mark]***.

2.6 As elderly patients are likely to have fewer functioning glomeruli than younger patients, fewer drug molecules will enter the Bowman's capsules every time blood passes through the kidneys ***[1 mark]***. This means that drug molecules will remain in the blood for longer, where they can continue to act at neuromuscular junctions and affect muscle contraction ***[1 mark]***.

2.7 If acetylcholinesterase was inhibited by anticholinesterases then more acetylcholine would remain in the synaptic cleft ***[1 mark]***, and be available to bind to receptors on the postsynaptic membrane/motor end plate to overcome the action of the neuromuscular blocker ***[1 mark]***.

3.1 The disease is caused by a gene carried on a chromosome that is not a sex chromosome ***[1 mark]*** and the person will have the disease phenotype even if they inherit just one copy of the disease allele ***[1 mark]***.

3.2 As this mutation is a substitution mutation, it will not cause the number of bases in the gene to change ***[1 mark]***. This means that the base triplets that follow the mutation will still be read in the same way, so it won't cause a frameshift ***[1 mark]***.

Remember, a frameshift can be caused by a deletion, addition or duplication mutation. These mutations cause the sequence of amino acids downstream to change.

3.3 The Generation 2 male is either heterozygous or homozygous for the healthy allele ***[1 mark]***. If the Generation 2 male is homozygous for the healthy allele then there is a 50% chance that the Generation 3 male will inherit the disease, meaning the student's conclusion is correct ***[1 mark]***. However, if the Generation 2 male is heterozygous then there is a 75% chance that the Generation 3 male will inherit the disease, meaning the student's conclusion is incorrect ***[1 mark]***.

It might be helpful to draw out a genetic diagram (e.g. a Punnett square) in your exam to help you get your head around what's going on.

3.4 Number of people in the city = 9.62×10^6 = 9 620 000.

Number of people in the city homozygous for the FH allele = 9 620 000 ÷ 1 000 000 = 9.62 = **10 people** ***[1 mark]***

Here you're working out the number of people that would be homozygous for the FH allele, so you need to round your answer to the nearest whole number (as you can't have 0.62 of a person).

3.5 $p^2 + 2pq + q^2 = 1$

Frequency of the homozygous dominant genotype = p^2

so $p^2 = 1$ in 1 000 000 $= 1 \div 1\,000\,000 = 1 \times 10^{-6}$

so $p = \sqrt{(1 \times 10^{-6})} = 0.001$

$p + q = 1$, so $q = 1 - p = 1 - 0.001 = 0.999$

Frequency of the heterozygous genotype = $2pq$

$= 2 \times 0.001 \times 0.999 =$ **0.001998**

[2 marks for the correct answer, otherwise 1 mark for recognising that the frequency of the heterozygous genotype = 2pq]

3.6 Inflammation of the β cells in the pancreatic islets of Langerhans could reduce the secretion of insulin into the blood ***[1 mark]***. Insulin normally lowers the blood glucose level by increasing uptake of glucose by cells ***[1 mark]*** and increasing the conversion of glucose into glycogen ***[1 mark]***. If insulin secretion is reduced then the blood glucose level can't be controlled properly/remains high, which could lead to Type II diabetes ***[1 mark]***.

3.7 E.g. a DNA fragment containing a copy of the normal LDL receptor allele / DNA fragments that would disrupt the mutated LDL receptor allele could be inserted into a vector ***[1 mark]***. The vector could then be inserted into the patient's cells ***[1 mark]***, where the DNA fragment(s) would become incorporated into the patient's DNA ***[1 mark]***. This would mean the patient produced the normal LDL receptor protein, which would increase the uptake of cholesterol from the blood ***[1 mark]***.

3.8 E.g. it could result in the overproduction of the LDL receptor protein. / The inserted DNA fragment could be incorporated into another gene / a regulatory region of a gene and disrupt gene expression elsewhere in the body. ***[1 mark]***

Answers

4.1 The increase in shade means that the crop plants may not have access to as much light as before, so their growth rate may be reduced ***[1 mark]***. This could mean that there is less food available for the rabbits ***[1 mark]***, so the environment can no longer support the same number of rabbits as before, meaning that the carrying capacity will have been reduced ***[1 mark]***.

4.2 Chlorophyll a or b can't absorb light beyond a wavelength of 700 nm, which could mean that ATP and reduced NADP aren't produced in the light-dependent reaction of photosynthesis ***[1 mark]***. This could mean that there is less ATP and reduced NADP available for the light-independent reaction/Calvin cycle ***[1 mark]***, so fewer products needed for plant growth (e.g. glucose) are produced ***[1 mark]***.

4.3 When far-red light is detected, TTA1 activity is no longer inhibited so the amount of indoleacetic acid/IAA produced will increase ***[1 mark]***. Indoleacetic acid/IAA stimulates the growth of plants shoots ***[1 mark]***, so an increase in the indoleacetic acid/IAA concentration could cause the plant to grow taller/bend further so that it is exposed to more light ***[1 mark]***.

4.4 Transcription factors control the rate of transcription of their target genes ***[1 mark]***. If two organisms with the same genome have different transcription factors acting on their genes then there will be different amounts of various proteins produced ***[1 mark]***, which will lead to the organisms having different phenotypes ***[1 mark]***.

5.1 50% / 0.5 / 1 in 2 ***[1 mark]***

5.2 As there are excess base triplets in the gene, there will be more than the usual number of amino acids in the protein that's produced ***[1 mark]***. This will mean that the protein can't fold into its unique tertiary shape, which will affect its function ***[1 mark]***.

5.3 Inhibitory neurotransmitters bind to receptors on the postsynaptic membrane at a synapse, which can cause hyperpolarisation ***[1 mark]***. This would prevent an action potential from being fired and therefore stop the transmission of a nerve impulse. / This may prevent excitatory neurotransmitters from being able to depolarise the membrane and trigger an action potential, which would stop the transmission of a nerve impulse ***[1 mark]***.

5.4 For: e.g. the person may want to know whether they carry the mutated allele before they decide whether or not to have children. / Knowing that they are likely to develop the disease may give the person time to prepare for the difficulties it may cause. ***[1 mark]*** Against: e.g. because there's currently no cure, knowing that they are likely to develop the disease may be difficult to cope with emotionally ***[1 mark]***.

5.5 The DNA is heated to 95 °C to break the hydrogen bonds between the two strands of DNA and therefore separate them ***[1 mark]***. This means that single DNA strands are created which act as templates to create the complementary strands of DNA ***[1 mark]***.

5.6 Electrophoresis separates out DNA fragments based on size, so copies of a *HTT* gene that have a larger number of CAG repeats would move a shorter distance than those that have a smaller number of CAG repeats ***[1 mark]***. Therefore, if the PCR products of the genetic test only move a short distance (e.g. compared to a control/sample taken from a person known not to have a mutated *HTT* gene) it suggests that the person is at risk of developing Huntington's disease ***[1 mark]***.

5.7 A small, double-stranded RNA molecule in the cytoplasm of a cell could unwind ***[1 mark]***. One strand of this RNA could then bind to the mRNA produced from the mutated *HTT* gene ***[1 mark]***. This could cause the mRNA to be cut into fragments / blocked ***[1 mark]***, which would stop the gene from being translated, meaning that the defective protein that causes Huntington's disease wouldn't be produced ***[1 mark]***.

6 ***21-25 marks:***
The answer includes material from a variety of different topic areas and clearly shows its link to the question title. No irrelevant material is included. The answer includes a range of detailed and accurate biological facts that are all of A-level standard. No incorrect material is included. Appropriate scientific terminology is used. Explanations are clear and the overall essay is very well written. (To get top marks, evidence of wider reading beyond the specification must be shown.)

16-20 marks:
The answer includes material from several relevant topic areas and links these to the question title. An irrelevant topic may be included. The answer includes a range of biological facts that are accurate and of A-level standard but may sometimes be lacking in detail. There may be one significant error in the scientific content. Appropriate scientific terminology is used. Explanations are clear.

11-15 marks:
The answer includes material from several relevant topic areas but doesn't link them to the question title. More than one irrelevant topic may be included. The biological facts included in the answer are mostly correct and of A-level standard but material is lacking in detail. There may be a few significant errors in the scientific content. Appropriate scientific terminology is usually used. Explanations are usually clear.

6-10 marks:
The answer includes material from one or two relevant topic areas but doesn't link them to the question title. Several irrelevant topic areas may be included. Some A-level content may be included but it will be lacking in detail and may contain several significant scientific errors. There may be limited use of scientific terminology. Explanations lack clarity.

1-5 marks:
The answer includes material that is only vaguely linked to the question title. Material is presented as a series of facts. Most of the material is irrelevant. The content is below A-level standard and contains a large number of scientific errors. Scientific terminology is not used or is below A-level standard. Explanations are poor or absent.

0 marks:
Nothing relevant is included in the answer or nothing has been written.

Here are some topic areas you might write about:

- the importance of ATP synthase in catalysing the formation of ATP from ADP and inorganic phosphate to allow energy to be stored and transported to cells (e.g. during photosynthesis/respiration);
- the importance of ATP hydrolase in breaking down ATP to release energy for energy-requiring reactions in cells (e.g. active transport across cell membranes/muscle contraction), and releasing inorganic phosphate for the phosphorylation of other compounds;
- the importance of DNA helicase in unwinding the DNA double-helix and of DNA polymerase in joining nucleotides in the new strand, during DNA replication;
- the importance of digestive enzymes such as amylases, lipases and peptidases in the hydrolysis of large food molecules into smaller molecules for absorption;
- the importance of the roles of rubisco and dehydrogenase enzymes in photosynthesis;
- the importance of enzymes involved in the control of blood glucose concentration (i.e. in glycogenesis, glycogenolysis and gluconeogenesis) including the importance of adenylate cyclase and protein kinase A in the second messenger model of adrenaline and glucagon action.

Remember, you should try to write about at least five different topic areas. You don't have to have written about only those listed here though — these are just to give you an idea. Make sure you link all the topic areas you have written about to the essay title. It's no good just writing about enzymes — you need to make it clear how they're important to living organisms.

Specification Map

This specification map tells you where each bit of the AQA specification is covered in this book.

Index

Index

Index

Index

BAR63